From Wild West to Modern Life
Semiconductor Industry Evolution

Dr. Walden Rhines

A SemiWiki.com Project

From Wild West to Modern Life: Semiconductor Industry Evolution

Copyright © 2019 by Walden C. Rhines

All rights reserved. No part of this work covered by the copyright herein may be reproduced, transmitted, stored, or used in any form or by any means graphic, electronic, or mechanical, including but not limited to photocopying, recording, scanning, taping, digitizing, web distribution, information networks, or information storage and retrieval systems, except as permitted under Section 107 or 108 of the 1976 US Copyright Act, without the prior written permission of the publisher.

Published by SemiWiki LLC
Danville, CA

Edited by Beth Martin

Although the authors and publisher have made every effort to ensure the accuracy and completeness of information contained in this book, we assume no responsibility for errors, inaccuracies, omissions, or any inconsistency herein.

First printing: April 2019
Printed in the United States of America

Contents

Foreword .... - 6 -
Building the Fundamentals .... - 8 -
   School Days .... - 8 -
   Choosing Stanford .... - 10 -
   The Story of HP-35 Calculator's LED Development and the Nobel Prize .... - 11 -
   Stanford and Semiconductors: A Unique Combination in the 1960s .... - 14 -
Learning at Texas Instruments aka the "Training Institute" .... - 17 -
   The Success of Morris Chang .... - 17 -
   TI Technology and Business .... - 18 -
   TI: Semiconductor Industry History of Innovation .... - 20 -
   Black Scholes and IC Design .... - 23 -
   TI Patent Priorities .... - 26 -
   Stubbornness Captures a Disruptive Technology and Leads to an Academy Award .... - 28 -
   Speak 'n Spell .... - 32 -
   Desperation Drives Inspiration .... - 36 -
   Why Do Brilliant People Like to Work Together? .... - 40 -
   Semiconductors Become a Worldwide Business .... - 42 -
   Apps Before there were Apps .... - 44 -
My View of EDA and Mentor from TI .... - 49 -
   EDA Grows: Systems Design vs. Integrated Circuit Design .... - 55 -
My View of EDA from the Top of Mentor Graphics .... - 59 -
   Developing Mentor's Strengths .... - 60 -
   Tales from Mentor CEO Seat—Avant! plays the Acquisition Game .... - 64 -
   Tales from Mentor CEO Seat—Carl Icahn Comes Knocking .... - 67 -
   EDA Cost and Pricing .... - 70 -
   EDA Market Dynamics .... - 70 -
   Honey, I Shrunk the EDA TAM .... - 74 -
   Basic Techniques for Managing an EDA Business .... - 77 -
References .... - 81 -

From Wild West to Modern Life: Semiconductor Industry Evolution

Foreword

In 1968, Texas Instruments, Motorola, and Fairchild dominated the emerging semiconductor business with 66% combined market share. Over the next fifty years, the industry de-consolidated – dozens of new semiconductor companies emerged, creating a more dynamic market that altered the list of the top ten largest companies.

During the same period, an ecosystem of companies emerged to grow the materials, develop the manufacturing equipment, design the software, and create all the other capabilities needed to support what has become one of the most strategic industries in the world. Much of this evolution was driven by relatively young, inexperienced individuals operating in a totally unregulated, free market, worldwide business environment. I was privileged to work with many of these people and to be involved in some of the revolutionary innovations.

Many people, including Daniel Nenni, have asked me to relate some of the stories of game-changing programs and people with whom I was involved, including the dynamics of growth of the Electronic Design Automation (EDA) industry. I've put this off for a long time, but Daniel is persistent. So I started writing some short vignettes during long airline flights. This activity required that I contact other people who were involved in this history, some of whom I hadn't seen for decades, to verify the accuracy of my recollections. I hope this collection of essays provides some feeling for the remarkable history of the growth of an industry as well as insights into its future evolution.

Walden Rhines
March, 2019

From Wild West to Modern Life: Semiconductor Industry Evolution

Building the Fundamentals

My path to the semiconductor industry was a winding one. Engineering was always to be in my future and I met a lot of interesting people along the way. Many of them from my time at Stanford for my Ph.D. studies, had either already made their mark on the development of semiconductors, or were destined to become leaders in the semiconductor industry.

School Days

Gainesville, Florida in the 1950s was a small university town of 25,000 people that doubled in size during the school year. There was almost no place to work except at the University of Florida so most of my peers had at least one Ph.D. parent on the faculty. The academic competition was fierce, as situations like the daughter of the head of the Math Department competing with the son of the head of the Physics Department for top scores in high school courses, raised the level of intensity.

My father was a professor of Materials Science and Engineering and a traditional engineer, as was his father. So, when it came to discussion of career choices, the conversation was short. "I think I might like to be a lawyer," I might say. "Engineering is great preparation for law school," my father would reply. If I suggested the medical profession, there would be a similar answer. Variants of this discussion were followed by more than a dozen visits to the leading engineering schools in the U.S. until he concluded that the University of Michigan had the best undergraduate engineering program. And so, in the fall of 1964, that's where I went.

While my father was pleased with my decision to affiliate with the Chemical/Metallurgical Engineering program, he was less enthusiastic about the love I acquired for computers. Michigan was among the first university recipients of an IBM 7090 mainframe and later an IBM 360 in the year I graduated.

University of Michigan's New IBM 360-67 computer, September 1968. Image source: https://aadl.org/N002_0682_003

Bruce Arden's Math 273 course attracted a lot of people I came to know, like Sam Fuller (later head of R&D at Digital Equipment and CTO at Analog Devices), David Liddle (founder of Metaphor Computer Systems), and Fred Gibbons (founder of the company that developed pfsWrite, the first widely accepted word processor for the TRS 80 and Apple computers). Math 273 required us to complete four computer projects including a program to execute the Newton/Raphson convergence approach to find functional values of zero for an equation. Little did I know that this basic algorithm would be fundamental to all the SPICE simulations I ran in the years ahead.

College wasn't all work, of course. Sam Fuller and I joined the Phi Kappa Tau fraternity and embarked upon various contests to test which had greater endurance, our brains or our livers. Holding multiple weekend parties and maintaining high grades in engineering were compatible only by repurposing the hours that we typically used for sleep. At some point, we decided that sleep was just an escape mechanism and unnecessary, so we tried to eliminate it. For the next year, I became susceptible to every cold, flu, or other illness that came along until, on the advice of a doctor, I found a way to resume a normal sleep schedule.

As graduation approached, the intensity of the Vietnam War increased. The 1967 Draft Act put an end to automatic deferments for graduate students unless they were married, which Sam was. Marriage struck me as much too extreme an alternative (since I didn't know anyone I wanted to marry), but I managed to find a program that let me go on to graduate school if I spent the summer at Fort Benning "pushing Georgia" with push-ups while my Drill Sargent yelled at me.

Choosing Stanford

Choosing a graduate school was a switch from my undergraduate experience. Getting into good graduate schools wasn't that difficult, so I only applied to Stanford and U.C. Berkeley. My father advised that "If you're good enough to go to graduate school, you're good enough to get someone else to pay for it." He lobbied for U.C. Berkeley as the real engineering school rather than the "science-oriented" Stanford. Graduate schools provided the funding back then, so the choice was mine.

Sam Fuller, who was number one in the Michigan Engineering class of 1968, was being recruited vigorously by MIT and Stanford. After many beer-laden discussions, we concluded that Stanford was the place to go. While Sam's wife finished her degree at the University of Maryland, Sam and I rented an apartment near the Stanford campus and entered the world of semiconductors and computers at Stanford in the fall of 1968.

Sam chose Ed McCluskey as an advisor. I chose Dave Stevenson, who had won a major DARPA contract to investigate III-V compounds. Dave's DARPA project also presented the perfect opportunity for a young engineering faculty member—a traditional metallurgist—to diversify into semiconductors. That professor, Craig Barrett, also joined my Ph.D. committee, along with Gerald Pearson, who had helped to develop the transistor at Bell Labs.

At 8:00 am on my first day of graduate school at Stanford, I joined the "Structure of Materials" class taught by Craig, the youngest faculty member in the Materials Science and Engineering Department. Craig had just returned from a post-doc in England and was energetically publishing papers, writing a book (along with Bill Nix and Alan Tetelman) and teaching classes. He passed out mimeographed copies (for a price) of the rough drafts of the book for the class textbook.

Craig's distinguished student career at Stanford led to a faculty appointment. His history at Stanford included a record in the high hurdles, which still stood from his undergraduate years. His relative youth as a professor had other tangible benefits for his students; he willingly joined us at the local watering hole, the "O" (short for Oasis, which continued as a student watering hole until it closed for good in March of 2018) and purchased pitchers of beer when the graduate student money ran out (which was early in the evening). Ultimately, his impatience with the academic world led to his departure to industry in 1974. He joined Intel and eventually became CEO (but that's another story).

Craig was the closest thing that Stanford had to an expert in electron microscopy, so he dutifully helped me analyze precipitates that were formed during the diffusion of zinc into GaAs to form light emitting diodes. I shared an office with Herb Maruska and we helped each other with our research. Herb's was focused on gallium nitride. This work resulted in the magnesium doped gallium nitride light emitting diode, which we patented in 1974, and which is the basis of all the blue LEDs today. The blue LED technology garnered the Nobel Prize in 2014 (although not for the Stanford research team).

Many of my predecessors in the Materials Science and Engineering Department at Stanford had worked on other aspects of III-V compounds, and some of them went to work at Hewlett Packard (HP) after completing their Ph.Ds. Although I have no way to verify its accuracy, it is their story that I retell here.

The Story of HP-35 Calculator's LED Development and the Nobel Prize

Hewlett Packard recognized that LEDs would be important for many types of instrumentation. The company pioneered LED research and eventually formed HP Associates to commercialize this business.

In the late 1960s, development began on what became the HP-35 calculator, the world's first scientific pocket calculator, introduced in 1972. The processor chipsets used in the production units were made by Mostek (which eventually became part of STMicroelectronics) and American Micro-systems Inc. Technology at that time made "reverse Polish notation" a more logical procedure for entering data; it reduces memory access and uses the stack to evaluate expressions. Many engineers still prefer the elegance of this approach.

The HP-35 calculator.

Choice of a display for the HP-35 logically fell to the most promising technology, GaAsP (gallium arsenide phosphide), which could be tailored to alter the exact wavelength of light emission. GaAs emits light at 1 micron, which is in the infrared and therefore not visible by humans. By alloying GaAs with phosphorus, the "band gap" can be tailored to emit at shorter wavelengths. Armed with this knowledge, the team of relatively young engineers attacked the development task of creating a suitable and reliable red LED for the HP-35 calculator. After analyzing various ratios of arsenic and phosphorus, they selected a combination that emitted a bright cherry red that was easily visible to the entire team of development engineers.

Detailed characterization and development of a manufacturing process followed. And then the day came to demonstrate their achievements. A presentation was put together for the Board of Directors of HP. The presentation included the multifaceted issues associated with light emission and culminated with a demonstration of an array of discrete LEDs that were arranged to spell the letters "HP."

When they pulled the switch, Bill Hewlett turned to David Packard and said, "I don't see anything." Packard agreed. Neither of them could see the cherry red LEDs.

What the young (under 40 years old) engineers had overlooked was the natural narrowing of bandwidth perception that occurs with age. Eyesight, hearing, smell, and almost all our senses deteriorate with age. As we age, the range of frequencies we can perceive decreases. The particular choice of GaAsP alloy that the younger engineers had selected was one that emitted red light in the visible spectrum (above 750nm). But visibility is relative. For those aged 60 or over, it wasn't visible. The entire project went back to the drawing board to reduce the wavelength of red light emission to a frequency that would be visible to a much broader range of the population.

Red wasn't the only LED color of interest at the time. A DARPA contract funded the work I was doing with GaAs, along with Shang-Yi Chiang, Craig Barrett and Herb Maruska. Herb had worked at RCA for James Tietjen and then Jacques Pankove[1] on gallium nitride before coming to Stanford on an RCA doctoral study award. They tried a wide variety of materials for LEDs so that RCA could build solid-state televisions. The challenge of short-wavelength emitters remained, making the lack of an efficient blue LED a limitation.

At Stanford, Herb tirelessly deposited thin films of GaN with various dopants. And then one day, we systematically analyzed his results. Element by element, I went through the periodic table while he told me who had tried various dopants and what the results had been. And then, miraculously, we focused on a group II element, magnesium, that was not well characterized with GaN. Herb headed to the lab and in a short period produced a film of Mg-doped GaN that emitted blue/violet light when a voltage was applied. We were all ecstatic and proceeded to apply for a patent with the help of the Stanford legal staff. The patent[2] was granted in 1974, and Herb returned to RCA a hero.

Unfortunately, the blue LEDs were not very efficient, so they were not ready for commercial production. RCA canceled the project in 1974. But, we published papers[3] and two researchers, Akasaki and Amano, talked to Herb about ten years later and were able to reproduce his results. The history is well documented.[4]

Later, the critical missing piece evolved. Shuji Nakamura of UC Santa Barbara fabricated Mg-doped InGaN LEDs that operated with a quantum well structure, dramatically improving the efficiency. Nakamura's advance was remarkable, and he clearly deserved the Nobel Prize that he received (along with Akasaki and Amano who were able to reproduce both Nakamura's and our results, as well as achieve stimulated emission). Nakamura highlighted the Stanford work when the Nobel Prize was announced, saying he believed recognition for the blue LED should also extend to Herbert Paul Maruska, for creating a functional blue LED prototype in 1972. Nakamura said he did not think his or Akasaki and Amano's work would have been possible without Maruska's contributions many years prior.[5]

Stanford and Semiconductors: A Unique Combination in the 1960s

Stanford engineering in the 1960s onward was full of many interesting people. The former Dean of Engineering, Frederick Terman, had recruited a variety of rising stars in the semiconductor industry including William Shockley and Gerald Pearson, both of Bell Labs transistor fame. Shockley was more famous because of the Nobel Prize and, ultimately, more INFAMOUS as he redirected his research from semiconductors to controversial theories of racial differences in intelligence. Since he had the office next to ours, we kept a sign in the window labeled "Shockley's Office is Next Door" just in case someone with a firebomb lost direction or became confused.

The McCullough Building was hardly a safe place anyway with research in II-VI and III-V semiconductors down the hall involving elemental materials that were poisonous in the parts per billion range. And T.J. Rodgers, who was later a founder of Cypress Semiconductor, was running experiments in the basement with the first of Stanford's ion implanters, causing unexpected and sometimes dangerous results.

While I plodded my way through Craig's "Structure of Materials" course, my extracurricular life was stimulated by my residence in Crothers Memorial Dorm, fondly referred to as "Cro Mem." It consisted of two buildings, side by side, one for graduate engineering and science majors and one for lawyers and MBA students. Although love was not great between the two buildings, there were frequent touch football games and mutual enjoyment of the promotional efforts of emerging wineries, like Wente Brothers and Inglenook, who provided free wine anytime we had a party, which was frequent. Judging from those I still know from Cro Mem, the wine promotion was very effective although maybe not for Wente and Inglenook.

But parties required more than wine, so we turned to the most innovative of the Cro Mem residents, Roger Melen. Roger arrived at Stanford with an undergraduate Electrical Engineering degree from Chico State. He published a book titled "Understanding Operational Amplifiers" (which I didn't) by his second year in graduate school, and he was making money in a variety of entrepreneurial ways, like consulting for Bay Area electronics startups or writing articles for Popular Electronics. Whenever we needed money for a party, Roger generously wrote an article, received $400 and the party was on.

Meanwhile, Roger worked on his Ph.D. thesis under Prof. Jim Meindl, who had dozens of graduate students (many of whom came to make up the Who's Who of the electronics industry) designing chips and producing them in the two-inch wafer fab on campus. Roger was working on the Optacon, a reading aid for the blind, developing an 8x16 pixel charge coupled device (CCD) for a compact version of the product. But he was much too innovative and productive to work only on his research. On the side, his consulting business was growing.

Roger designed the electronics for all sorts of equipment that recent graduates were developing. Since most of these companies had very limited cash flow, Roger had to be content to accept future royalties in payment for some of his work. Over time, he discovered that these entrepreneurial customers, although skilled in engineering and product development, had "lost the ability to count." Roger was concerned about being cheated on the royalties, so he developed a system to overcome this deficiency. He performed the design work as usual, but instead of labeling the integrated circuits (ICs) in the design, he and his graduate student friend, Harry Garland, re-marked them with proprietary letters for one of the key ICs in each design. He then assumed the supply chain fulfillment role by relabeling the ICs and providing the parts for production.

Roger and Harry created a company named Cromemco, after the "Cro Mem" dorm, which in a few years became one of the very early, successful microprocessor-based computers. They needed funding and publicity to start a company, so Roger turned to his tried and true technique—writing articles for Popular Electronics, but this time under the Cromemco name.

As I made my way to the end of my research in 1972, my committee member Gerald Pearson told me not to worry about finding a job. He would take care of it. Sure enough, Professor Pearson was good to his word. When I told him I was ready, he picked up the phone and called one of his former graduate students named Morris Chang (who was at Texas Instruments and later went on to found TSMC). Most Stanford Ph.D.s in my field at that time remained in the Bay Area to work for Fairchild, National Semiconductor, Hewlett Packard, or other local companies. But Texas Instruments (TI) was the largest semiconductor company and TI recruited heavily from the Stanford Ph.D. crowd. Professor Pearson also connected my lab partner, Shang-Yi Chiang, who later became head of R&D for TSMC. Shang-Yi and I both went to TI to begin our careers in the semiconductor world.

Learning at Texas Instruments aka the "Training Institute"

I completed my degree and headed to Texas Instruments (TI) in Dallas in 1972. My first project was developing CCD imagers. You can imagine our shock at TI when we saw a cover article of Popular Electronics entitled "Build Your Own Solid State Imager" by my classmate Roger Melen along with his graduate friends Terry Walker and Harry Garland. While Fairchild, Sony, RCA, and TI competed fiercely to develop early CCD imagers, it looked like the graduate students at Stanford had beaten us to the punch. Or so we thought.

The article provided circuit diagrams plus a block labeled "solid state imager." To fill that block, the article instructed the reader to send a check or money order for $25 to Cromemco to buy the needed component. But instead of sending a CCD imager, Roger sent American Microsystems S4008-9 DRAMs. This early DRAM did not automatically refresh the bits during readout and came in a ceramic package with a metal lid that could be replaced by a quartz lid by popping the tops off the ceramic packages. The image quality was good enough for the hobbyists. Those $25 checks added up to over $50,000 and became critical seed money for Cromemco, which Roger and Harry sold in 1987 to an electronics company called Dynatech, who was a major user of Cromemco systems for the display of weather forecasts.

The Success of Morris Chang

About a month after I arrived at TI, Morris Chang was promoted to Vice President of the Semiconductor Group, leading to an association that I value to this day, more than 45 years later. Morris was born in mainland China, near Shanghai, and had the distinction of being accepted to Harvard University in the U.S. Sensing that a technical degree was the best path forward, Morris transferred from Harvard to MIT after the first year. He earned a Mechanical Engineering degree from MIT and took a job at Sylvania, an early participant in the emerging semiconductor industry.

After several years of industry experience, Morris was attracted to TI in 1958 and was hired to develop and manufacture transistors for IBM's first major mainframe computer with transistor logic, the IBM 7090. Four transistors were produced by both IBM and TI, three of which were yielding at acceptable levels.

Yield in semiconductors means the number of units manufactured that actually work. But one of them remained at low single-digit yields. Morris worked late nights analyzing data and finally figured out the problem. Yields soared and he became a hero, ultimately becoming the manager of all of TI's germanium transistor business.

Morris' goal, however, was to become vice president of R&D. His superiors told him that such a goal would be impossible unless he had a Ph.D. since most of the researchers had that credential. So Morris took advantage of a TI-funded opportunity to go to Stanford in 1961. He studied under John Moll, Bill Spicer, and Gerald Pearson, receiving his Ph.D. in record time in early 1964.

When Morris returned to TI, the business had grown dramatically, so instead of joining the research labs, they asked him to run the integrated circuit business. Although TI's Jack Kilby invented the integrated circuit in 1958, it's market position slipped when Bob Noyce at Fairchild Semiconductor developed the planar process in 1960. TI lost to Fairchild in the first bipolar generation resistor-transistor logic (RTL) and the second generation with diode-transistor logic (DTL). But, by making a few good decisions and putting in a lot of hard work, TI emerged as the leader in bipolar transistor-transistor logic (TTL) integrated circuits, a technology developed by Sylvania. This TI success, combined with TI's two year lead in developing the silicon transistor took TI to its goal of $1 billion of revenue. This was the situation at TI shortly after Morris returned; it had become the world's largest semiconductor business.

TI Technology and Business
While TI dominated the bipolar semiconductor era of integrated circuits and had the largest market share in the semiconductor industry in the 1960s, the MOS era that evolved in the late 1960s led by Intel and Mostek was a different story. TI did the best job of the "Big Three" (TI, Fairchild, and Motorola) of making the transition from bipolar to MOS technology initially. But when the so-called "Hogan's Heroes," the gang of seven under the Chairman and CEO, Lester Hogan, left Motorola en mass, they were replaced by a substantial number of TI's senior semiconductor managers in the mid-1970s including Jim Fiebiger, whose team from TI changed the competitive environment. MOS memory and

later microprocessors became strengths for both of TI's key competitors, Intel and Motorola. That made the late 1970s and 1980s a difficult period for TI.

Intel's 1103 1K DRAM became a widely adopted standard. TI had three programs to match it before a production worthy part was developed, but it was too late to catch up. Hope appeared when the 4K NMOS DRAM emerged since Intel had a three-transistor cell and TI leapfrogged to a single-transistor cell, but the victory was short-lived. Mostek had introduced an undistinguished 4096 MNOS structure (silicon nitride/silicon oxide gate) for their 4K product but, upon hiring Paul Schroeder from Bell Labs (who said on his resume that he was the greatest DRAM designer in the world), usurped the lead with the Mostek NMOS 4027. Meanwhile, TI struggled with its TMS 4030 design and remained allied with the camp of companies doing 18- and 22-pin parts because of the advantage that they required no multiplexing of address and data and would, therefore, be faster than the 16-pin Mostek part. Mostek's 4027 disproved that thesis. The only solution for TI was to copy the Mostek 4027, which was perfectly legal at the time, and I was chosen to head that team.

I moved from Dallas to Houston to begin that project. On the day I arrived in Houston, Dick Gossen, head of memory design, advised me to begin filling out my resume. Dick explained to me that the corporate senior management was underestimating the difficulty created by the analog nature of a DRAM. TI and everyone else had freely second-sourced logic parts by copying. In fact, that was the normal procedure that was encouraged in the industry to make a design viable, i.e., solicit another company to copy your part and help to make it a standard. DRAMs, however, have analog sense amplifiers that have variable behavior depending upon the process used for manufacturing.

I began a detailed analysis of the device structure of the Mostek 4027 and discovered that when the Mostek founders left TI to start Mostek, they took the TI process with them. So the normal difficulties of matching a design and a process were not relevant; the process was the same.

We followed the 4K with a similar copy of the 16K DRAM that kept TI as a contender in the MOS memory business, but the challenges of the quickly evolving microprocessor business put TI further behind[6].

The industry was at a major turning point, and it is rare for a company to remain in a leadership position through two major transitions of an industry. TI had led with the silicon transistor and the bipolar integrated circuit. But the next two generations of MOS memory and MOS microprocessors left TI behind. Of course, TI still exists today—the only company to continuously hold a spot in the top ten largest semiconductor companies from the 1950s to the present—partly because the tide was reversed in the next generation of embedded DSPs.

TI: Semiconductor Industry History of Innovation

Texas Instruments is a remarkable company founded by remarkable people, and Eric Jonsson—TI co-founder, president, and then mayor of Dallas—was one of the most remarkable visionaries of the 20th century. He was a renaissance man who created an industry and a fortune by following the needs of the emerging oil exploration industry, then semiconductors. He followed his tenure at TI by serving as mayor of Dallas starting in 1964. Over his three terms, he took the city over from the depression of being the site of the Kennedy assassination to being one of the most innovative centers of commerce in the 21st century. He co-founded an educational institution that became the University of Texas, Dallas. Today, Dallas is home to more than 10,000 corporate headquarters, including over twenty Fortune 500 companies. But it didn't happen by accident.

The roots of TI go back to northern New Jersey in 1930. J. C. "Doc" Karcher developed reflection seismography technology that could be used to reveal the character of strata beneath the earth and to predict the most likely places to drill for oil. With a $500K loan from Everette Degolyar, chairman of Amerada Oil, Karcher founded Geophysical Services Inc. (GSI). The East Texas oil field moved the center of momentum for the oil exploration industry to Texas and so in the 1930s, GSI moved with it. GSI filled in excess exploration crew time creating their own seismic database for a subsidiary Coronado Oil. When GSI's business and Coronado's conflicted, the more valuable company, Coronado, was sold to Stanolind and the GSI management, Eric Jonsson, Eugene McDermott, Cecil Green, and Bates Peacock, took out a loan to buy the GSI remainder on December 6, 1941. Since there was no significant shortage of oil during World War II, Jonsson applied GSI's skills in electronic equipment to win a government contract for magnetic anomaly detectors for submarine patrol planes. The Navy contracting officer was Pat Haggerty, who later joined GSI in 1945 and ultimately became Chairman and CEO of Texas Instruments.

The seismic exploration side of the business continued to grow after the war. One unusual success was a sole source contract with ARAMCO for exploration of the Arabian Peninsula. ARAMCO's contract with Saudi Arabia required them to turn back a block of land every five years to Saudi Arabia, so they hired GSI to figure out which blocks of land were LEAST likely to contain petroleum reserves. The electronic equipment side of the business grew much faster than the seismic part, resulting in total revenue in 1951 of $7.5M of which seismic contracts were less than $3M. Disagreements about the focus of the company led to a buyout of the company by Jonsson, Haggerty, McDermott, and Green as well as a new name, Texas Instruments.

When AT&T offered licenses for transistor technology in 1952, TI was not initially invited to the licensing meeting. TI eventually found a supporter in Bell Labs and was able to send Haggerty and Mark Shepherd to that meeting to purchase a license for the transistor technology.

Not trusting their success to luck, TI also hired Gordon Teal who had been a primary developer of the techniques to grow single germanium crystals to make the transistor possible. Surprisingly, TI emerged as a key contender in the race to produce transistors for military and commercial use. Even better, the TI team was the first to successfully produce silicon transistors, which provided much better thermal stability than germanium. They announced the achievement at NAECON in May 1953. During this same time, TI developed germanium transistors for radios. When none of the radio producers would buy the product, they turned to IDEA corporation and co-developed and introduced the very successful Regency radio in 1954. This success producing high volumes of transistors was fundamental to winning IBM's business in 1958 for the first transistorized computers. This and other successes, plus an acquisition, took the revenue to $300M in 1963.

During the summer of 1958, Jack Kilby worked through the TI summer shutdown because he was a new employee and had accrued no vacation time. He spent his time creating a phase shift oscillator on a single chip with transistors connected by gold wires, bonding the discrete devices together on the same chip. Just like that, Kilby created the integrated circuit. Old timers I met at TI told me that it had been obvious that you could put more than one transistor on the same piece of silicon. "But why would you want to do it?" they asked. "You would never get both of them working at the same time." Today, billions of transistors work in harmony to solve problems, but it wasn't so obvious then.

It was an obvious move to Jack, however, and the subsequent litigation over the invention of the integrated circuit led to one of the most significant patent lawsuits of history, enhancing the career for Roger Borovoy, corporate counsel for Fairchild.

Jack insisted in his testimony that the words "laid down" applied to deposited metal electrical connections and the patent stated that, "Electrically conducting material such as gold may then be laid down on the insulating material to make the necessary electrical circuit connections." But TI hired a prestigious law firm that thought the patent suit would be a slam dunk for TI and didn't do much preparation. It wasn't.

Borovoy prevailed with the view that the planar process that was dated five months after the TI invention was distinctly different from Jack's approach to connecting the elements of the integrated circuit. Borovoy moved on to Intel and became well recognized as a corporate attorney. Jack had to accept the failure of the TI-chosen law firm and share the recognition of invention of the integrated circuit with Robert Noyce, although Noyce's premature death made Jack the sole recipient of the 2000 Nobel Prize in Physics for the integrated circuit. Sharing the recognition was not a totally negative outcome. While Jack's words "laid down" included the planar process in his view, the compromise to recognize both men settled the West Coast/Dallas dispute and brought us all together—a result that Jack, as a gentle non-argumentative person, would have applauded.

I had the good fortune to meet regularly with Jack at TI and later. He was a truly wonderful person, very modest and quiet. We both joined the advisory board of FormFactor, Inc., at the request of investor Bill Davidow, where Jack and I had many delightful discussions.

Black Scholes and IC Design

From the earliest days of my childhood, I was always trying to find ways to make money—paper routes, lawn mowing, coke sales at football games—I did it all. And, except for a motorcycle I bought during junior high school when, at age 14, I could get a driver's license in Florida, I saved most of the money. During high school, I bought my first publicly traded stock, Eastman Kodak, and fortuitously profited from the introduction of the Kodak Instamatic Camera six months later, instilling me with the dangerous idea that I had some sort of intuition for investments despite the random nature of the luck.

So it should be no surprise that as I worked on challenging research projects in TI's Central Research Laboratory, I also became deeply involved in trading standardized stock options when the Chicago Board Options Exchange was created during my first year at TI. Pretty soon I was doing "butterfly spreads," "ratio writes," and even selling "naked calls."

The TI SR-52 programmable calculator.

My trading activity stepped into high gear with the introduction of the SR-52 programmable calculator, as TI tried to catch up with the HP 65 programmable calculator that was already in the market. I went to work writing programs to improve returns and reduce risk in my stock option investing program.

Not long before this, Fisher Black and Myron Scholes published an article in the "Journal of Political Economy" providing a mathematical derivation to calculate the intrinsic value of a stock option. Myron Scholes later won the Nobel Prize and became a director and principal of a company, Long Term Capital Management, which experienced a blowup so big that Alan Greenspan writes about the threat it posed for worldwide financial stability in his book, "The Age of Turbulence: Adventures in a New World." I went to work implementing the Black Scholes formula on the SR-52. The formula is a complex equation, so it required some effort to squeeze it into the limited memory of the SR-52.[7]

Volatility data was not generally available for most stocks, so my use of the Black Scholes model focused on comparisons of options with different strike prices and expiration dates, where the volatility assumed in the equation would be constant. And then I began using it for trading. My broker at Merrill Lynch became fascinated, and soon many of the brokers in his office had SR-52s.

One day, I became aware of a request from the management of the Professional Calculator Department at TI for sample programs written for the SR-52 that could be used as examples to attract customers, especially for applications other than engineering. I went to a meeting and met Robb Wilmot and Peter Bonfield (now SIR Peter Bonfield, who I've known ever since). They were excited by my options trading program and decided to run a full page ad in the Wall Street Journal offering customers a free copy of the program. It was a big success, and I seriously began considering a career move into financial analysis software.

As Steve Jobs said in his commencement address at Stanford[8], connecting the dots that will be important to your career is difficult looking forward. In this case, the connection with Robb and Peter in the Calculator Products Division, or CPD, had an interesting consequence. Later that year, a decision was made to move CPD to Lubbock, Texas because the division was growing so fast that space needs couldn't be accommodated in Dallas. For people like Robb and Peter, who came from the UK, both Dallas and Lubbock were near the edge of civilization, so they could easily adapt to the new environment in Lubbock. But for most of the employees in Dallas, a move to Lubbock didn't sound attractive. Lots of management slots opened up, including the job of Engineering Manager for the division, supervising 150 engineers who designed the chips and developed tooling for the plastic case injection molds for calculators.

I am told that someone in the Calculator Division suggested, "That guy who wrote the Black Scholes program – wasn't he some type of chip design manager in the Central Research Lab? I wonder if he would be willing to move to Lubbock?" And that's all it took. A few weeks later, I inherited responsibility for a group of people who had to be convinced that moving to Lubbock would be a good experience.

Most amazing was the group of managers who agreed to move. Those of us reporting to Ron Ritchie, the Division VP, included:

- Robb Wilmot – Later became CEO of ICL, International Computers Ltd., the largest computer company in Europe
- Peter Bonfield – Later became CEO of ICL, then CEO of British Telecom and subsequently served on boards including TSMC, Astra Zeneca, Ericsson, Sony and nine other public companies including Mentor Graphics. He has 11 honorary degrees and is currently in the news because he is Chairman of the Board of NXP. He is now Sir Peter.
- Tommy George – Later became CEO of Motorola Semiconductor
- Kirk Pond – Later became CEO of Fairchild Semiconductor
- Jim Clardy – Later became CEO of Harris Semiconductor and then CEO of Crystal Semiconductor, which became Cirrus Logic

Following is a slide (referred to as a "Vu-Foil" back then) of the agenda for part of the 1978 Strategic Planning Conference that TI held each year. P. Pfeiffer refers to Eckhard Pfeiffer, who later became CEO of Compaq and M. Chang refers to Morris Chang.

CONSUMER ELECTRONICS BUSINESS OBJECTIVE
1978 STRATEGIC PLANNING CONFERENCE

TUESDAY, MARCH 21

3:15	CONSUMER ELECTRONICS BUSINESS OBJECTIVE	R. RITCHIE	15 MIN.
3:30	P&AE STRATEGY	T. GEORGE	15
3:45	WORLDWIDE MARKET DEVELOPMENT	J. MIGLIORE	15
4:00	EUROPEAN THRUST	E. PFEIFFER	15
● 4:15	TECHNOLOGY	W. RHINES	15
4:30	PROFESSIONAL CALCULATORS	P. BONFIELD	15
4:45	HANDHELD PRODUCTS	K. POND	15
5:00	PRINTER/DISPLAY CALCULATORS	J. CLARDY	15
5:15	TIME PRODUCTS	R. SHELLY	15
5:30	PERSONAL COMPUTERS	R. NADEN	20
5:50	NEW BUSINESS OPPORTUNITIES	G. HELMS	15
6:05	CRITIQUE	M. CHANG	15

A stellar group of technology managers.

TI Patent Priorities

Probably the most innovative person I met at Texas Instruments, other than Jack Kilby, was Ken Bean. Ken had a list of patents that would impress even the most skeptical. He started his career at Eagle Picher and came to TI in the mid-1960s. He was a warm, delightful and modest person but very innovative when it came to finding solutions for silicon manufacturing problems. He worked in Semiconductor Group product divisions as well as research labs over his TI career, as did Mike Cochran, a topic that I'll address later.

Ken rarely saw a semiconductor manufacturing problem that he couldn't solve. When TI had problems introducing the "thermal printer" that was used in the "Silent 700," Ken had a solution that made the silicon print heads manufacturable. One of the most innovative patents that Ken filed was the patent on the slicing of silicon wafers. Easy, don't you think? No. Ken addressed a problem for DUF (or diffusion under film) in bipolar integrated circuits. "Pattern shift" was a problem that occurred because early bipolar integrated circuits used wafers that were oriented to <111> or <110> crystal planes.

As a result, subsequent layers of deposition "shifted" modestly as the epitaxial layer grew in the direction of crystal orientation. This caused a shift in the alignment of subsequent photomasks. Not a problem for Ken. He was called in to solve the problem, and he did. "Why not slice the wafers a few degrees off the perfect orientation," Ken asked. "Then the DUF layer wouldn't follow the crystalline orientation." It worked. Subsequently, wafers for bipolar integrated circuits were sliced slightly away from perfect orientation.

In the early 1970s, Monsanto decided to get out of the semiconductor wafer business and showed up at TI with a list of patents for which they hoped to claim royalties (since TI still manufactured its own polysilicon and silicon wafers). After Monsanto showed their patents, TI lawyers passed Ken's patent[9] to them, showing why wafers used for bipolar semiconductors are sliced a few degrees away from the perfect orientation. The story goes that the Monsanto lawyers looked at the patent and closed their briefcases. That was the last that the TI lawyers saw of them. It was truly a fundamental patent in the early days of semiconductor history. I loved my interaction with Ken, and he loved our family. We kept in touch until his death, he kept our Christmas cards on his refrigerator, and he delighted in the success that TI ultimately achieved.

The HP Silent 700 portable terminal.

One of the things that Ken taught me was the importance of customer interaction in the innovation process. Ken had assignments in the Semiconductor Group and in the Central Research Labs as well as the Semiconductor Research and Development Lab. Interestingly, he generated patents at approximately the same rate per year regardless of where he was working. The same was true of Mike Cochran, who worked in a variety of organizations in TI, including both semiconductor product groups and research laboratories (and is partially responsible for the Cochran-Boone patents on the microprocessor).

I decided to analyze the patent productivity of the truly great patent generators like Ken and Mike. Fortunately, TI had a system that helped me. After the TI DRAM lawsuits, TI management realized that patents were a very important source of royalty revenue, much to the dismay of many TI engineers including me, who had been taught that patents should only be used defensively, to allow TI to enter new markets. TI created a special segment of the annual performance review process that rewarded the creators of the most valuable patents. Those lawyers who negotiated the patent cross-licenses voted on the most valuable patents. The result: I now had a list of the most "valuable" patents.

The result of the analysis amazed me, although I wasn't allowed to publish the results. But the conclusion was clear. People like Ken Bean and Mike Cochran generated about the same number of patents per year. But the ones that they generated when they were in product groups turned out to be much more valuable than those they generated when they worked in research organizations. Why? I concluded that, because the patents they filed when they were in product groups were developed in response to a customer problem, they grew in value as more competitors adopted similar solutions to the same type of problems. The other patents sounded great; they just weren't as valuable because they were generated by innovative ideas rather than customer problem-solving.

Stubbornness Captures a Disruptive Technology and Leads to an Academy Award

Back when I first joined TI in 1972, I was assigned to work on a new contract that had just been awarded and badly needed staffing. The U.S. Department of Defense had decided that solid-state charge-coupled device (CCD) image sensors were going to be a strategic technology and they formed a joint services program under Larry Sumney (who later became CEO of the Semiconductor Research Corporation, serving in that position for more than thirty years).

Fairchild had hired Gil Amelio from Bell Labs and was promoting buried channel technology because of its high efficiency, i.e., by using ion implantation to shift the minimum of the electrical potential to store charge below the surface of the silicon, any losses due to surface state interactions were minimal. Meanwhile, RCA was a clear contender in this emerging business because of their experience with video cameras and associated technology (as was Sony, but as a foreign company, Sony could not be funded by the U.S. DoD).

TI was desperate to be included in the contract shootout, so they proposed a totally different approach—building the CCD on a silicon wafer and then thinning the devices from the back side to about a 25-micron thickness. This approach avoided the losses associated with shining light on the front side of the device where polysilicon and metal interconnect interfered with light transmission. The CCD thinning technology was relatively simple. We used wafers with a 25-micron thick, lightly doped p-type epitaxial layer on top of a heavily p+ doped substrate. The p-layer served as an etch stop leaving the 25-micron paper-thin layer. TI's proposal looked good to the Navy's Night Vision Lab for use in "Starlight Scopes" and, at the last minute, TI was added to the contract. Today, virtually all solid-state imagers are illuminated from the back side of the silicon but that approach really didn't take off for thirty more years.

Meanwhile, Dean Collins, who ran the CCD Imaging Branch, was able to promote the technology to other branches of the government, generating lots of additional funding for TI. One particularly difficult contract called for building a moving target indicator that would store and compare successive images. Larry Hornbeck took on the task but stubbornly refused to fabricate the device as originally proposed. Instead, he pursued what he called a stratified channel CCD architecture, the first-ever CCD to have the capability for storing two overlying charge-storage and transport channels. With the assistance of people like Ken Bean for the epitaxial process development and Jerry Hynecek (inventor of TI's virtual-phase CCD) for the modeling task, Larry proved the concept with backside illumination of thinned, packaged devices.

Dean sold another program to the DoD, this time for a "solid state light modulator" that again relied on the thinning expertise we developed. It used a hybrid manufacturing process to produce a frontside, deformable mirror spatial light modulator, with backside CCD-addressing. The deformable mirror was a continuous sheet of a metalized polymer membrane. Once again, Larry came up with a different approach, consistent with his habit of taking on management to pursue approaches that ultimately proved to be superior and more manufacturable on the path to his "Digital Micromirror Device" or DMD.

Larry was convinced that creating arrays of individually-addressable cantilever micromirrors along with a monolithic manufacturing process would solve the problems of defects in the array and lead to much improved optical performance. By this time, I had taken on the job of President of the Data Systems Group. Tom Stringfellow, who managed the Peripheral Products Division of my Group, began funding Ed Nelson to support a potentially revolutionary approach to printing using the digital micromirrors.

George Heilmeier, who was one of the first senior TI managers to be hired from outside the company, became VP of Research for TI and supported Larry for the chip development. By 1986, Larry had developed and patented the first practical methods for manufacturing high-density arrays of micromirrors on an integrated circuit in a conventional wafer fab. This IP served as a barrier to competitors who would have immediately started developing their own version of the digital micromirror device once it was publicly disclosed in 1988. Thirty-one years after the invention of the digital micromirror device in 1987, TI is still the only manufacturer of this disruptive technology, a highly unusual, and possibly unique, example in semiconductor history.

Larry made a pivotal decision in 1987 to attach the micromirrors to torsional suspensions and actuate them into contact with rotation stops. This made it possible to manipulate light with the precision of time division by pulse-width modulation, increased the optical efficiency and reduced the address voltage. And so, the digital micromirror device (also DMD) was born.[10] The DMD became commercially known as the digital light processing (DLP) chip.

The tiny mirrors of this device assumed a "1" or "0" position and pulse width modulation was applied to control the pixel intensity, a method that required extensive algorithmic development to produce high-quality projected images. Numerous other problems had to be solved, such as the gradual increase in surface stiction to the point where a mirror would stick in a "1" or "0" position, a problem Larry immediately addressed with a novel, electro-mechanical release mechanism and in 1990 with a surface treatment that he developed (despite his claim that he hated chemistry).

Through all this, the Semiconductor Group didn't want to take the product to production. Potential applications, like projection TV, would require major investments with questionable business benefit since the light modulator component was a small part of the total system cost. But TI, with some government help, provided enough funding to keep it alive and Larry's persistence provided the momentum.

And then, in 1985, Jerry Junkins became TI's CEO. Jerry was looking for a semiconductor project with system implications that could make a real difference to the company. Jerry's background was running the defense business of TI and the DMD looked good to him. So, he redirected the staff of an older (four inch) wafer fab that was destined to be shut down and totally re-dedicated it to working out the bugs in the DLP chip.

TI built a total portfolio of know-how, software, CMOS-based manufacturing technology, and intellectual property to lock up an amazingly disruptive technology. The entire motion picture industry distribution structure was totally changed as its 115-year-old projection systems were replaced by "DLP Cinema" technology and software-based distribution of motion pictures. TI approached other applications including printing and home projection systems with a solutions approach that included the basic DLP component, algorithmic development, manufacturing, and application engineering. The revenue approached $1B, and companies like Samsung and RCA introduced televisions based upon the DLP because of its extremely bright, sharp colors.

Ultimately, Larry Hornbeck, the innovator and developer of the technology, was nominated for and received an Academy Award of Merit (Oscar statuette) in 2015. Most interesting to me, however, is the fact that TI kept the DLP program alive for almost twenty years before any real revenue was realized. This wasn't the only example of persistence at TI, accompanied by tolerance for inflexible innovators, and it is part of the reason that TI is the only semiconductor company that has ranked among the ten largest since the 1950s.

Larry Hornbeck receives an Academy Award of Merit.

Speak 'n Spell

Success has many authors, and the Speak 'n Spell product from TI demonstrates this. For most of the semiconductor industry, results of innovation were not apparent to the masses but, for the consumer electronics that emerged in the 1970s, the innovations were visible, exciting and fun. My job in the Consumer Products Group (CPG) was Engineering Manager with responsibility for the design and development of all the chips and plastic cases used in TI's fledgling consumer business.

In early 1977, almost all of CPG was moved from Dallas to Lubbock. From then on, we performed the logic design for our chips in Lubbock while the physical layout was done in Houston under the direction of Krishna Balasubramanian (Bala), an energetic, driving manager who was perfect for the task of juggling dozens of complex designs while competing for resources with TI's traditional semiconductor business.

From left to right, Gene Frantz, Richard Wiggins, Paul Breedlove, and Larry Brantingham.

Paul Breedlove was in charge of product development for the Consumer Calculator Division, which was managed by Jim Clardy (later CEO of Harris Semiconductor and then co-founder and CEO of Crystal Semiconductor, which ultimately became Cirrus Logic). Paul and Jim had a miserable job. Japanese manufacturers like Casio, Sharp, Toshiba and many more could design and manufacture great "four-function" (add, subtract, multiply, and divide) calculators for less than TI could. By late 1977, TI was reselling Toshiba four-function calculators with a TI label because they were more profitable than our own. Paul kept searching for a differentiating alternative, and he found it by attending one of the monthly "Research Review" meetings held in the Central Research Laboratories and open to TI employees from other parts of the company.

At this particular review, Richard Wiggins presented the technology he was developing for speech synthesis. He was approaching a capability of producing understandable speech at a data rate of only 1000 bits per second. Paul was fascinated. Why not develop a product that took advantage of speech to differentiate, or augment traditional consumer electronic products? Paul was helped along by the analogy of one of the few really profitable, successful consumer calculators called The Little Professor, which was an arithmetic learning aid for children. Every year we expected revenue for the Little Professor to decline, but it seemed to have a life of its own. We were beginning to realize that parents will pay any price to give their children an advantage in the education system.

As an experiment in innovation, TI had recently established a funding mechanism called the Idea Program where any employee could propose an idea for a product or technology and, if approved, receive $25,000 of funding to demonstrate feasibility. Paul submitted an Idea Program proposal (probably because the Consumer Calculator Division was really squeezed for funding) and Ralph Dosher, the CPG Controller, approved it. That's when I became involved. Paul needed someone to figure out how to design chips that could be used in the product. Larry Brantingham worked in the Logic Design Branch of the Engineering Department I ran, and he became the obvious choice.

Speech synthesis chips were under development at National Semiconductor and other companies but success was very limited because the current state of the art N-Channel MOS, or NMOS, technology was just too slow to achieve the needed performance for this computationally intense application. What is so remarkable about Speak 'n Spell is that Larry didn't use the higher performance NMOS technology but instead used the much slower P-Channel MOS, or PMOS. Why, you might ask? Very simple. Larry didn't know how to design with NMOS. In addition, the Consumer Products Group (CPG) was in a continuous battle with the Semiconductor Group over the pricing of chips. Morris, then the Semiconductor Group VP, became tired of all the arguments and settled the dispute by offering CPG a flat $25 price per three-inch wafer of PMOS, which was a five photomask process at that time. If Larry had learned how to design with NMOS, the program would have failed because the cost of NMOS wafers would have been too high. While the artificially subsidized price for PMOS made the cost feasible, the performance seemed much too slow.

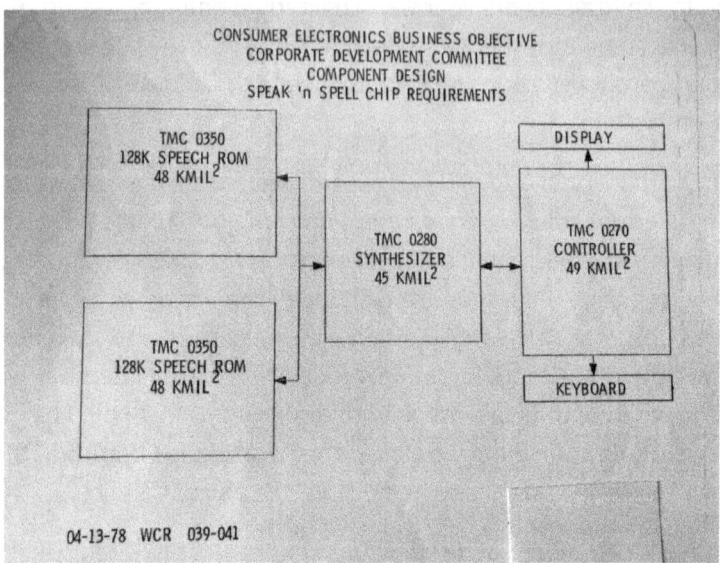

Vu-foil slide I presented to the Corporate Development Committee in April 1978 to show our approach to the Speak 'n Spell chip designs.

Larry went to work with Richard Wiggins designing a pipelined multiplier in PMOS. Responsibility for the product was moved from Consumer Calculators to Specialty Products which was run by Kirk Pond (later CEO of Fairchild Semiconductor) because, although Consumer Calculators was struggling, the Specialty Products Division was struggling even more to find new product possibilities. Gene Frantz was enamored with the product and quickly became a product manager, tasked with all the issues of choosing product features, name, marketing, etc. as well as managing the overall product development.

Even more ridiculous than designing a pipelined multiplier for the synthesis chip was the task of designing read-only memory (ROM) chips big enough to store the pre-recorded speech vocabulary. When I presented the proposed chip design program to the TI Corporate Development Committee, Dean Toombs, head of R&D for the Semiconductor Group, argued that the engineers in CPG had gone crazy. Semiconductor Group was struggling to produce the NMOS 2716 ROM at very low yields. If they couldn't produce a 16K ROM, how could

CPG design a 128K bit one? What Dean overlooked, however, was the fact that the 16K bit ROM in the Semiconductor Group needed an access time of 450 nanoseconds while our speech chips could be dramatically slower. PMOS was also an older, more mature technology and was easier to produce than the highly advanced NMOS. We got approval to go ahead and received corporate funding to develop a four-chip system with a synthesizer, controller and two, 128K bit ROMs.

As the actual die size increased beyond the original estimates, the estimated cost of the chips increased. At one point, Kirk Pond threatened to kill the whole program because it was well known that $40 was a critical point for consumer products where both spouses had to approve the purchase. By the time Speak 'n Spell was introduced (marketed as Speak & Spell), the suggested retail price was $60, and it sold so well that we quickly raised it to $65. As with the Little Professor, parents just couldn't resist the purchase of an educational aid that would help their children spell, even though the synthesized speech sounded more like a robot than a human.

Shortly after the introduction, we were invited to show the product on the Today Show, the most popular TV news program of the day. Charley Clough, our highly articulate and lovable head of Semiconductor Group sales walked Jane Pauley through the steps of using Speak & Spell while Gene Frantz was behind the stage with backup units since the reliability of our early production units wasn't very good.

Speak & Spell took the world by storm and became a great story of corporate innovation. Not long after the introduction, I moved to Houston and took over the Microprocessor Division where we developed the TMS 320 single-chip digital signal processor (DSP). Although I was the only one who worked on both programs, there was at least a remote connection to the theme of DSP in TI's speech synthesis success, and DSP became the cornerstone of TI's next wave of growth.

Desperation Drives Inspiration

1978 was a bad year for TI. In April, Intel announced the 8086 followed by disclosures of 16-bit microprocessors from Motorola, the 68000, and Zilog, the Z8000. TI had tried to leapfrog the microprocessor business by introducing the

TMS 9900 16-bit microprocessor in 1976. But the TMS 9900 had only 16 bits of logical address space and the industry needed a 16-bit microprocessor for address space rather than for performance. In addition, TI had no peripheral chips for the TMS 9900 and tried to overcome that weakness with an 8-bit bus version of the 9900 called the 9980 (an approach that Intel also followed with the Intel 8088). TI found that any performance advantages of a 16-bit microprocessor were sacrificed with the 8-bit approach.[11] Intel overcame that weakness by winning the design socket for the IBM PC with the 8088 despite the performance weakness.

As TI engineers tried to develop a 16-bit TMS 9940 microcontroller, they discovered a whole new set of problems, resulting in resignation or termination of much of the microprocessor team. I became manager of the TI microprocessor activity more because nobody wanted the job than through personal merit. But I had a different motivation.

At the time, I was engineering manager of TI's Consumer Products Group, heading the design of calculator chips, Speak 'n Spell speech processors, and other miscellaneous devices. That job was located in Lubbock, Texas, which was not my idea of a great location for a 31-year-old single male. So Houston, which had some drawbacks, scored far above Lubbock in my plan. Most of my time in Houston was initially filled by exit interviews with all the people who were bailing out of the sinking ship. Fortunately, there were some resilient, smart people like Kevin McDonough, John Hughes, Jeff Bellay and Jerry Rogers (who later founded Cyrix, a successful floating-point processor and X86-compatible microprocessor company, and married Jodi Shelton, Founder, and CEO of the Global Semiconductor Alliance). John Hughes convened a day-long meeting to debate what would be important after host microprocessors since we had obviously lost that race.

The answer: Special-purpose microprocessors. We chose three and then added a fourth one later, and named them the TMS 320, 340, 360, and 380. The TMS 320 was a communications processor, the 340 graphics, the 360 mass storage. Later the TMS 380 was designed for the IBM Token Ring LAN. The first job was to decide what a communications processor, or Signal Processing Microcomputer, as we called it, would be.

Ed Caudel spent the next six months analyzing that question and concluded that the distinguishing characteristic was a single-cycle multiply/accumulate instruction (although we required two cycles in the first generation TMS 32010 but made it to one cycle with the 32020). John Hughes commissioned Kevin McDonough and others from systems groups around the company to write applications using alternative instruction sets. Early on, we found we needed a DSP expert and, fortuitously, our group in Bedford, England had interviewed one named Surendar Magar. Tony Leigh has documented most of the history very accurately. Magar quickly determined that the single cycle multiply/accumulate would have to be done in hardware, not software as Ed had hoped.[12]

TI was not the first company to develop a single-chip DSP. In fact, it was the fifth. Intel announced one while we were developing the TMS 320 but it incorporated an on-chip 5-bit A/D and D/A, making it unusable for most applications. Chi-Foon Chan, Co-CEO of Synopsys, who was working at Intel on the first DSPs, tells me that the poor customer reception of the 2920 caused Intel to kill the enhanced version of the chip, which he was working on, thus keeping the door open for TI.

Despite lots of delays, the TMS 320 was announced at the February 1982 ISSCC with rave revues from people like Ben Rosen, the leading semiconductor analyst. We knew we had a winner but the world didn't understand digital signal processing. We had to publish books, develop algorithm libraries, and promote the technology. Financial analysts paid no attention and neither did our senior management so I found myself giving largely unappreciated presentations at financial and technical meetings as well as in the TI board room.

We needed some high-volume applications but our largest customer was Lear Sigler who was making analog repeaters for underwater cables. Hardly a high volume application. We needed consumer products companies in Asia. But our Japanese organization was totally uninterested. Their customers almost always wanted custom chip designs. And then a unique event changed the tide. A group in Canada wrote an application note on how to design a FAX MODEM using a TMS 32010. A group in Australia read the article and built a prototype and sold the design to a Japanese company, Murata.

A Murata engineering manager called the TI Japan office and asked for a quote on the TMS 32010. The TI Product Marketing Engineer had never heard of the TMS 320 but he looked it up in the price book and quoted a $35 price. We had never sold one north of $10 so this was a unique response. The Murata engineer said, "Good. I'll take 20,000 parts." From then on, we had no resistance from the TI Japan organization and, in fact, they then designed a derivative named the TMS 320C25, which became one of the highest-volume members of the family.

The most strategic success came later. After years of struggle, we convinced Ericsson to design a TMS 320 into a cell phone. A subsequent need for a lower-cost version of the phone became apparent. We had to combine two ASICs, a TMS 320 DSP, and a static RAM into a single chip. "How hard can this be?" I said. All the parts are already verified. I didn't understand the laws of verification that drive the need to verify internal state, increasing the amount of verification as the square of the number of gates when you combine chips. I willingly committed to Lars Ramquist, the CEO of Ericsson, that we would do the design quickly. A crash effort resulted and Gilles Delfassy took on a similar task for Nokia.

Fortunately, the chips worked and TI grew the wireless baseband MODEM business to something approaching $4 billion per year. The subsequent step was even more critical. To do similar low-cost designs for all the other producers of cell phones (and other applications like hard disc drive controllers), we needed to combine our ASIC library with our embedded DSP.

Everyone told me that this would be suicide. ASICs were sold on a cents-per-gate basis, while DSPs had high gross margins. But Bala and I decided to combine the ASIC and microprocessor business into one group under Rich Templeton. A good decision. Success followed, DSP-based technology became nearly half of TI's revenue, and Rich eventually became Chairman and CEO. In between, Tom Engibous leveraged the technology to create a wide variety of businesses while building TI's position in analog. In 2017, TI became the most profitable of the major semiconductor companies in the world at 41% operating profit.

Why Do Brilliant People Like to Work Together?

In high technology, there are numerous instances of highly productive groups coming together and generating game-changing ideas and products. This happened at Shockley Semiconductor in the 1960s when Gordon Moore, Bob Noyce, Jean Hoerni and more found each other and took advantage of Sherman Fairchild's offer to start a semiconductor company. It also happened to me in Houston, Texas in 1978 (a much less likely place than Palo Alto, California).

TI had a late start in the microprocessor contest, focused its attention on calculator chips, and was left behind by Intel and Motorola in the general-purpose host microprocessor business. But failure has a way of stimulating the desperation needed for success, and the group in Houston went on to develop the TMS 320, the first really successful single-chip DSP, and a host of other important technologies.

Although TI arguably has the original microprocessor patent (awarded to Mike Cochran and Gary Boone), the MOS Division was struggling just to produce MOS memory. The Microprocessor Group was focused on a strategy that would catch up with Intel by second-sourcing the Intel 8080A, develop TI's own set of 8-bit microprocessors and peripherals (the 5500 series), and then leapfrog with the TMS 9900 16-bit chip that would also be used by the computer and defense businesses of TI.

Brilliant junior designers like Kevin McDonough and Karl Guttag were involved in the process when I arrived in October 1978. The group was in melt-down mode because the 16-bit microcontroller, the TMS 9940, was in its sixth or seventh re-spin and looked like it would never work. Although good engineers were resigning at a rapid rate, we had just formed a new group in Bedford, England. This was the first case I know of where design teams were organized around the world to do 24-hour-per-day design. We had groups of engineers assigned to a particular product in Japan, England, and the U.S. who could, if needed, pick up the work of each other as the sun moved around the globe and the databases remained in our IBM 4341 computers.

The Bedford, England design group was assigned the task of developing peripheral chips for the TMS 9900 16-bit microprocessor. The most notable was the TMS 9914 which implemented the HP GPIB standard. The chip became a long term success despite the lack of success for the TMS 9900. The team even anticipated the risk that others would copy their chip, so they went to great lengths to disguise the transistors, making enhancement mode devices look like depletion mode, just to confuse anyone who tried to copy.

A small group was assigned responsibility to develop the TMS 9918 graphics chip for the TI Home Computer.[13] While the TI 99/4 Home Computer was a disaster, the chip was not. It led to the development of new concepts in graphics and became part of a standard known as MSX that was promoted by Kazuhiko "Kay" Nishi, CEO of ASCII Microsoft, and was used by more than twenty different computer and video game manufacturers. Many people in graphics development are still familiar with the term "sprites," a graphical representation that was developed by the TMS 9918 team. This same group went on to develop the TMS 340 graphics processor that was adopted by IBM for the 8514A standard that, unfortunately, experienced a short life before being replaced by VGA in the IBM PC.

About a year after I arrived in Houston (from my previous job as Engineering Manager of Consumer Products in Lubbock), we combined all the logic design resources in Houston under one manager, Jerry Rogers. Jerry had been a career enlisted man in the Navy and joined TI after retirement as a technician while he worked on his engineering degree at the University of Houston. He was an effective manager but very tough, with no sympathy for any performance less than the best. He had a thick skin and was willing to push back on management, a trait that helped with many successes.

During this period in Houston, we hired an amazing array of innovative engineers. TI started a program to train new sales application engineers by assigning them to short stints in the product divisions. Rich Templeton was one of those early assignees. We liked him so well that we convinced him to join our group and give up the rotational training program and he did. Later he became Chairman and CEO of TI. K. Bala was his supervisor. One day in about 1991, Bala mentioned in a conversation with me that he thought one of his employees might be his future supervisor. "Who is that?", I asked. "Rich Templeton, and I think he might be your boss as well," said Bala.

Over the years, people who started their careers in that group in Houston eventually managed much of the company. We needed a marketing manager for DSPs when David French (later CEO of Cirrus Logic) was running the business, so we brought in Mike Hames who was in the Bipolar PROM group and knew nothing about DSP. When French left to join Don Brooks at Fairchild, we brought in John Scarisbrick to manage the DSP business, and he later took it to new heights.

One of the most impressive capabilities came when we needed improved manufacturing. Yukio Sakamoto, the most capable operations manager I've ever known, joined us to run all the manufacturing operations. He was dissatisfied with our status, and so he promoted Kevin Ritchie multiple levels to the job of DMOS 4 Wafer Fab Manager. People tell me that Ritchie became one of the most effective manufacturing managers in the semiconductor industry and recently retired after a distinguished TI career as Senior VP of Technology and Manufacturing. Sakamoto became CEO of Elpida Memory, the company that combined NEC's and Hitachi's DRAM businesses.

Semiconductors Become a Worldwide Business

Among the companies that bought a license from Bell Labs to produce the transistor was Sony. While the U.S. maintained its lead in technology, other countries like Japan emerged as competitors. Semiconductor manufacturing was both labor intensive and capital intensive. Fairchild became the first major semiconductor manufacturer to start operations overseas, adding an assembly site in Hong Kong in 1964 where labor costs would be lower.

TI and Motorola followed, although TI began with a misstep by starting an assembly site in Curacao. TI made up for this slow start through a different path—an attempt to sell in the Japan market. After World War II, U.S. companies were not allowed to set up wholly owned subsidiaries in Japan; they had to partner with a Japanese company who would have majority ownership. Companies like IBM and Kodak that had operations in Japan before WWII were grandfathered and could continue with their 100% owned subsidiaries in Japan.

TI wasn't interested in a joint venture. Pat Haggerty saw the potential that Japan offered as a future manufacturing powerhouse. So this became the first case of TI using its U.S. patent portfolio for reasons other than defense. The negotiations resulted in permission from the Japan government allowing TI to set up a joint venture with Sony in 1968 merely for appearances. I'm told that Sony people never showed up and TI quietly bought out their share of the business later. But Haggerty established a relationship with Morita, founder and CEO of Sony.

TI began a successful offshore assembly and later wafer fab operation in Hatogaya, Japan on the outskirts of Tokyo, followed by another assembly site in Hiji Japan on the island of Kyushu. The Hiji site was on the top of a small mountain overlooking the ocean on three sides and must have been one of the most valuable pieces of industrial real estate outside Tokyo. This habit of finding valuable real estate for plants was a TI characteristic, rumored to be the responsibility of Board member Buddy Harris. The choice of the TI plant in Nice, France was terrible from the point of view of location for manufacturing but it was on the top of a hill with a breathtaking panoramic view of the French Riviera.

Soon the race for offshore manufacturing sites was on. Morris Chang's influence came to bear and Taiwan would have been the next site, but Morris tells me that the Taiwanese government wasn't flexible enough. TI, therefore, built the Singapore site in 1968, then Taiwan in 1969, Malaysia in 1972 (simultaneously with Motorola and SGS Thomson in Kuala Lumpur) and the Philippines in 1979 (a site that I was proud to have report to me starting in 1987).

TI did two things that were unique among semiconductor companies in the race to build up offshore manufacturing. First, TI decided that cheap labor was not the only reason to go offshore. The offshore sites had skilled technicians as well. So TI moved automated manufacturing equipment to its offshore sites even though manual labor was cheap. This turned out to be highly advantageous. The other unique thing TI did was to establish wafer fab manufacturing in Asia, starting in Japan.

Intel remained largely in the U.S., Motorola was primarily the U.S. and Europe as were most other semiconductor companies. Europe was necessary, at least for assembly, because they had substantial duties on imported semiconductors. European assembly sites saved money despite the high labor cost. TI, of course, had wafer fabs all over Europe, starting in the UK, then Germany, France, and Italy. Assembly sites were limited to Portugal and Italy.

One result of the establishment of wafer fabs in Japan was a creation of awareness of the superb manufacturing process variability control that was possible with Japanese workers. In cases where we sent the same photomask set to Japan, the die sort yields were typically much higher than those of the same devices produced in the U.S., which TI used to its advantage.

When the trade wars between Japan and U.S. semiconductor companies erupted in the 1980s, MITI (the Japan Ministry of International Trade and Industry) assigned Japanese companies quotas for the purchase of semiconductors from U.S. companies. Sony was assigned a very high quota of 20%. All the Japanese companies wanted to fill their quotas with DRAMs, but only TI and Micron were still in the business in the U.S. At this time, I was managing an organization I named Application Specific Products, or ASP, that had responsibility for microprocessors and ASICs. Yukio Sakamoto and I went to Japan to negotiate a deal with Sony with a goal of having TI manufacture the chips used in the industry standard Sony Walkman.

Because of the historical relationship between TI and Sony, my meeting started with Norio Oga, the CEO and former opera singer who succeeded Morita as Sony's CEO. Sony's offer: If you can match the Sony Semiconductor internal transfer price and quality, you can have 100% of the business. When we started production, our packaging cost alone for an 84-pin Quad Flat Pack was six cents per pin, more than the total price of the chip plus package. Within four months, thanks to Sakamoto, we were at one yen per pin. Similar ratios existed for the chip. And over the next year, we billed Sony for $200 million for Walkman chips and greatly enhanced our manufacturing capability.

Apps Before there were Apps

My development of a calculator program to determine the Black Scholes value for an option was not the only application that attracted financial people to programmable calculators. As the SR-52, and later TI 59, grew in popularity and took market share from the HP 65, we began to discover a vast community of innovative people writing programs for these calculators.

Peter Bonfield and Stavros Prodromou drove the formation of PPX-52 Professional Program Exchange, a forum where a contributor of a useful, well-documented program could receive credits for the purchase of other programs. As these programs accumulated, TI moved to publish booklets of programs on various topics and sold them. Because of the success of my Black Scholes program, and because we were short of person power, I was appointed to provide management supervision for the PPX Xchange.

Each month, we met to review the new programs that had been submitted. It was at that point that I began to comprehend the enormous resource available to us. Thousands, maybe millions, of talented people wanted to demonstrate their expertise through programmable calculator programs. In most cases, they didn't care if they were compensated. They just wanted to show other people how brilliant they were.

The ultimate example came when I reviewed a program for a "one-armed bandit" that simulated a Las Vegas slot machine. By loading the program and then pressing "enter," the display showed three single-digit random numbers separated by dashes. Dashes? There were no dashes on the SR 52 or TI 59 calculators. How could they possibly have done this? It took one of our expert engineers to analyze the program and figure out how it worked. The creator of the program had discovered that execution loops could be created that would simultaneously display more than one number; since the segments in the LED displays were strobed at about 14 times per second, the program could create overlap and thus a dash between numbers. The program was so brilliant that we contacted the author to see if he wanted to work for TI. Similarly, the SR 52 had extra, undocumented registers that programmers discovered and used for applications that were not anticipated by the developers of the SR 52.

Over the next year, many communities of people became connected through their common interest in different applications. TI's published booklets of applications carried contact information for the authors of programs. Although there was no internet for authors to communicate, they found ways to share information. TI then sponsored events to showcase the diverse set of applications available for the programmable calculators.

From Wild West to Modern Life: Semiconductor Industry Evolution

I was asked to demonstrate my Black Scholes program at one of these events in New York City where analysts and others from the financial community were targeted. Ben Rosen, the Morgan Stanley semiconductor analyst who was the most respected in the industry, came to the event. Ben was fascinated with the Black Scholes program and invited me to tour the trading floor at Morgan Stanley. Later, he visited Lubbock, Texas (a major trip for a New York investment banker) and we showed him what we were doing. I continued to run into Ben at conferences and other events.

And then, after I moved to Houston to run the Microprocessor Division, I received a strange phone call. It was Ben. He said that he was leaving Morgan Stanley and that he and L.J. Sevin were starting a venture capital fund. And he said that he would be in Houston the following week and wondered whether I would be able to have dinner with him and L.J. Of course, I was available.

Ben gave his pitch for how he and L.J. wanted to set up potential entrepreneurs and fund them while they worked on ideas for new businesses. That way, any conflict of interest with present employers could be avoided. I was flattered that they would think of me. In fact, I was amazed that they would make a trip to Houston just to talk with me. A few months later it became apparent that they had not come to Houston just to talk with me, as evidenced by the announcement that Sevin Rosen would fund Compaq Computer, a Houston startup headed by Rod Canion, another TI employee.

I didn't take advantage of Ben and L.J.'s offer. My responsibilities were growing too rapidly at the time to consider leaving TI. But since Ben had to sell rights to his semiconductor newsletter, the Rosen Electronics Letter, Esther Dyson bought it and renamed it Release 1.0 and began the reorientation from semiconductors to software. She, along with George Gilder, continued some of the semiconductor theme. When TI announced the TMS 7000 8-bit microcontroller, I made a trip to New York and did a series of one-hour interviews with representatives of various electronics journals.

Gordie Campbell, then CEO of SEEQ, gave the presentation with me as our alternate source for the TMS 7000. Gordie highlighted the Ethernet controller that SEEQ had embedded in their version of the TMS 7000. After giving the presentation seven times during the day, Gordie and I became bored and we switched places; I gave his presentation and he gave mine. And that's how we met Esther, who wrote up the announcement in her newsletter and then proceeded to invite us to speak at the PC Forum each year.

My View of EDA and Mentor from TI

During 1980 and 1981, while I was at TI, three computer automated design companies were founded: Daisy, Mentor, and Valid. Daisy and Valid offered custom hardware workstations plus software to provide the unique capabilities required by engineers. Mentor took a different tack in not creating custom hardware, which turns out to have been a critical decision for its future. Charlie Sorgy at Mentor evaluated the Motorola 68K-based Apollo workstation just being introduced and concluded it could provide everything that Mentor needed. Meanwhile, Jerry Langler and others worked on the software product definition, interacting with design engineers to zero in on a set of capabilities that would solve the design problems of those designing electronic chips and systems.

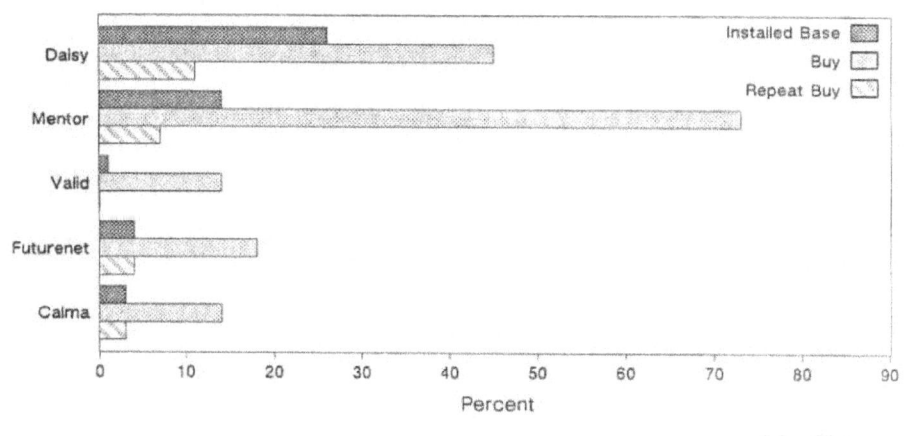

EDA Market data from 1986.

Until this time, semiconductor companies developed their own design tools, typically running on large mainframes. In the 1970s, that meant the largest IBM mainframes available. The mainframes at most corporations were shared with the rest of the company. The result: not much design work was done during the last week of the quarter when the corporation was closing its books because the computers were loaded to capacity.

At TI, we considered our in-house EDA software to be a competitive differentiator. Much of the success of TTL for TI came from the ability to crank out one design per week with automated mask generation with the "MIGS" system. Other semiconductor companies had their own EDA capability. It was somewhat revolutionary when Calma introduced its automated layout system at about the same time as Computervision and Applicon did, all based upon 32-bit minicomputers. Computervision, Calma, and Applicon became the big three of the first generation layout tools.

TI considered these newcomers as a threat and continued to develop a system based on the TI 990 minicomputer. Meanwhile, TI's competitors were quickly adopting Calma and, in some cases, Applicon systems. Design engineers were rebelling when I arrived in Houston in 1978. We were using the TI 990 based "Designer Terminal" and Ramtek displays to do the layout editing with a light pen. Designs were digitized on home-grown systems based on layouts that were created by draftspersons (mostly draftsmen) on a grid matched to the design rules of the chip. Our people wanted to buy a Calma system, and so we did. However, just as we got our first Calma workstation, GE acquired Calma and quickly lost many of the key employees. With the introduction of the Motorola 68K in early 1979, a host of companies, including Apollo, began developing a new generation of engineering workstations.

By now, the management of TI's Design Automation Division (DAD) realized the limitations of its approach. The 1982 Design Automation Conference in Las Vegas, where Mentor introduced its first product called IdeaStation, further affirmed the need to move to a next-generation approach. TI became one of the first major semiconductor companies to commit to the newly introduced Mentor IdeaStation product, which was based on the Apollo workstation. The Apollo workstation provided clearly superior capabilities to our minicomputer system. TI signed up for a complete conversion to Mentor's system, becoming, for a while, the largest company user site for Apollo computer systems, with more than 900 workstations in use at the peak.

The Apollo DN400 workstation.

Internal support groups in large corporations don't usually surrender their corporate roles, despite their competitive disadvantage, and TI was no exception. A plan existed to complement the Mentor software with TI's DAD-developed software. Mentor readily agreed since TI was a very large customer win. TI's management accepted the whole strategy because of the strong history of success of DAD in maintaining a competitive design advantage for integrated circuit design.

But TI did not end up being a good customer for Mentor. The Mentor software had been adopted by much of the military/aerospace and automotive industries, which needed standardization of design capture and simulation processes across their companies. Mentor had realized major success with systems companies in aerospace, defense, and automotive industries and was rapidly becoming a worldwide standard, especially in Europe. The DAD engineers were experts in design software and they wanted to modify the software, customize capabilities and do all sorts of things that distracted Mentor from its strategic direction.

Meanwhile, my role in the TI Semiconductor Group changed. I was appointed President of the TI Data Systems Group, TI's $700 million revenue business in minicomputers and portable terminals and was moved to Austin, Texas. While I was away over the next three years from late 1984 through mid-1987, a decision was made to divorce TI from Mentor. So TI and Mentor parted ways. TI ported its proprietary software to the Apollo workstation, and Mentor focused on supporting the needs of systems companies.

In mid-1987, I returned to Dallas as Executive VP of the Semiconductor Group only to find that the original move to commercially available design automation products had been reversed. TI was once again an island in an industry that was building on broad innovation from a diverse set of designers working with commercial EDA suppliers. Unlike the early semiconductor history when TI had nearly 40% market share, TI now had no more than 10% share and the economies of scale didn't justify custom design tools.

One of the entities that reported to me initially as EVP of the Semiconductor Group was DAD. I appointed Kevin McDonough to assess our position in design automation and recommend a solution. His conclusion: Adopt the design automation platform from Silicon Compiler Systems Corp. (SCS) and move to "RTL-based" design. And so we did. We committed $25 million to SCS and started a conversion of the MOS microprocessor portion of our design business to SCS. Few people even remember that SCS, which was an outgrowth of an AT&T Bell Labs spin-out called Silicon Design Labs, or SDL, actually developed the entire language based top-down design methodology before VHDL and Verilog even existed. TI and Motorola were among the first adopters.

Securing the TI account gave SCS the credibility for acquisition, in 1990, by Mentor Graphics at a premium price. Why would Mentor acquire SCS when they already had a strong IC Station product that could effectively compete against Cadence Virtuoso predecessor products? The answer: a top-down language methodology was clearly the direction of the future for the semiconductor industry. The problem: The two methodologies (top-down, language-based versus detailed layout) were disruptively different. Traditional designers viewed RTL design as the province of "computer programmers." The traditional, i.e., "real," IC designers knew how transistors worked and could craft superior ICs with a detailed design and layout system.

What evolved was internecine warfare. Hal Alles, VP of IC Design at Mentor, a veteran genius developer from Bell Labs and founder of SDL, had the undesirable challenge of convincing the two groups to work together. They didn't. Step by step, the SCS designers denigrated the traditional IC design approach and Mentor's message was bifurcated. The result: a window for Cadence to become the clear leader in traditional IC detailed design.

Meanwhile, other companies exploited the fact that SCS had a closed system for language-based design using two languages, L and M, which were proprietary. VHDL and, much later, Verilog, became public domain languages for top-down design.

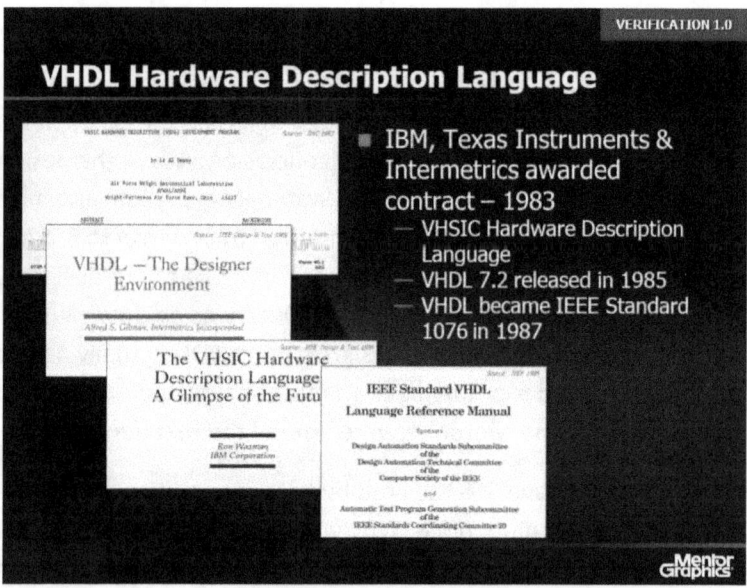

The development of VHDL.

At TI from 1982 through 1984, we initiated a research proposal to create an open language for top-down design called VHDL, or VHSIC Hardware Description Language. In 1983, we were granted a contract to develop VHDL with IBM and Intermetrics as co-developers. I was one of five speakers at a special event to announce the plan in 1983. In 1987, VHDL became an IEEE standard 1076. In

1985, Prabu Goel formed a company, Gateway Automation, that subsequently developed an even simpler language called Verilog. The company was acquired by Cadence.

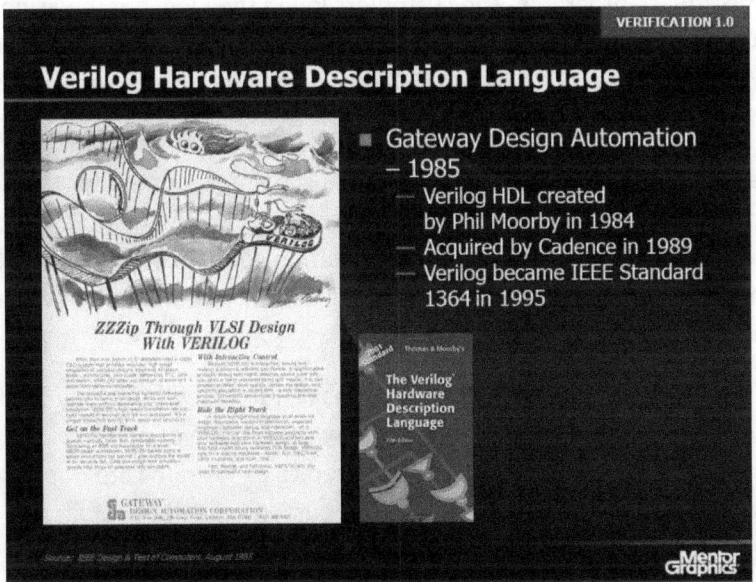

Verilog joins in the hardware language party.

The final result: Engineers who were accustomed to schematic capture were gradually displaced by language-based developers. The EDA industry went into one of its major discontinuities, the transition from schematic capture to RTL-based design. Mentor lost a lot of momentum and SCS never really became a standard for RTL-based design, although it might have been if it had made its languages open.

During this time, semiconductor companies like TI became more important to the EDA industry, which had focused more on systems companies. Most semiconductor companies didn't have TI's history of EDA software development, so a commercial EDA industry became increasingly viable. As the growth of the EDA industry accelerated, it became apparent that a new generation of product was needed. There were no standards for user interfaces for engineering workstations from Apollo, Sun, or other new competitors. Interoperability with third-party applications was not well supported.

Mentor decided to focus on solving the problem of interoperability with a new version of their tool they called Version 8.0 (Falcon). Falcon was to be a unified environment that could utilize the same user interface and database across different software programs. The customers loved the idea, but it was not a good approach from a technical standpoint. One more example of Clayton Christianson's "Innovator's Dilemma" where doing what customers say they want is frequently not the right solution.

In retrospect, it is clear that a single database is not appropriate for the wide variety of formats required for integrated circuit design. By the time Falcon was in development, starting in 1988, the workstation companies had solved the problem of standard user interfaces themselves, so ultimately there was no need for Mentor to provide one.

Mentor's most critical mistake, however, was terminating the legacy family of design software without completion of a new generation of products. As the schedule for Version 8.0 slipped (later referred to as Version Late dot Slow), there was less and less product available for customers to buy. Mentor's revenue peaked at about $450 million and declined to $325 million with lots of employee frustration and resignations as the entire company was mobilized to save Version 8.0.

The Falcon approach never really worked. Mentor went through a very difficult period because of the failure of Version 8.0. Rarely does a software company go into decline and then recover. The reason is that software companies have a large fixed-cost base of employees and, when their revenue declines, they have no choice but to reduce personnel, which makes recovery difficult. By 1991, Mentor was in deep trouble.

EDA Grows: Systems Design vs. Integrated Circuit Design

EDA began and grew with the integrated circuit (IC) design business probably because IC design grew in complexity faster than printed circuit boards. The race for superiority in PCB design evolved in parallel, however, and has become increasingly important as system design moves to more advanced EDA.

Daisy, Mentor, and Valid supported a combination of IC and PCB design. Both technologies required schematic capture and layout, but simulation was primarily an IC design technology. Mentor and Daisy targeted both IC and PCB design while Valid specialized in PCB. At the same time, companies like Racal Redac (Europe), Cadence, SciCards (on VAX), Intergraph and others competed for the PCB market. As in much as the IC market, competitive advantage in PCB design and layout (and eventually manufacturing) resulted from strategic acquisitions as well as organic technology development.

Computervision, Calma, and Applicon were the "Big Three" electronic design environments that preceded the Daisy, Mentor, Valid era. But the GE acquisition of Calma, which had a very strong IC layout capability, demonstrated how large companies could easily mismanage the acquisition of fast-moving, small, high-tech companies, and the value of Calma was quickly lost.

Daisy and Mentor went head-to-head and Mentor ultimately won the majority of the systems companies (and even owns the remnants of Daisy today through Mentor's acquisition of the PCB design tool Veribest from Intergraph in 1999). Winning the systems companies is an important accomplishment that gives Mentor its continued strength in system design today. Systems companies (particularly for aerospace, defense, and automotive) rarely changed their preferred EDA suppliers, even as they adopted IC design tools to complement their PCB tools.

A critical shift occurred in the early 1990s. Mentor's PCB capability came from the acquisition of California Automated Design, Inc. (CADI) in 1983, Mentor's first acquisition. Cadence had acquired tools as well, and both Zuken and Racal Redac had strong positions grown from organically developed tools. In 1990, Cadence and Mentor had approximately equal market shares, with Zuken and Racal Redac making up much of the remainder of the PCB market.

Then, in 1991, Cadence made a very bold move, taking advantage of the fact that Mentor was in a period of weakness due to its struggles with Version 8.0, which had been in development since 1988 and was still not released. Cadence acquired Valid, announcing that the overlap between Cadence and Valid PCB design tools would be quickly resolved by eliminating the losers and crowning the winners.

This turned out to be a difficult strategy since ALL of the users from both Cadence and Valid lost some portion of their design flow. That forced all the Cadence and Valid users into a competitive re-evaluation of all the alternatives. Zuken gained a little and Mentor gained a lot, while Cadence kept some. The result: By 1999, Mentor had 20% of the PCB market, Cadence 17%, and Zuken, who had acquired Racal Redac to complement its Japan strength with a European supplier, had 16%. By this time, the dot com crash was beginning and Zuken reduced investment while Cadence focused on IC design. Mentor, who was still troubled by the Version 8.0 problems, continued a heavy rate of investment in "system design" including PCB, as an area of #1 market strength, and continued to gain market share in PCB, peaking at a market share of about double its nearest competitor.

Over the next two decades, all this history had an effect on strategic evolution. The original companies that adopted EDA standardization in the 1980s were largely systems companies. They needed standardization in design methodologies, libraries, and tools across their disparate divisions. Even though two-thirds of Mentor's revenue ultimately came from IC design, the original adopters of EDA remained as a stable base of customers, particularly those who manufactured cars, planes, and trains, or were involved in aerospace and defense. Mentor was able to capitalize on that large market share and, thanks to some developments along the way, developed a leading position in electronic wiring and embedded software for those kinds of systems.

As much as anything, this systems capability is what made Mentor so attractive to Siemens' software division as they looked to extend their "digital twin" platform from design, product life cycle management, mechanical CAD, and manufacturing simulation to the electronic dimension of the digital twin.

From Wild West to Modern Life: Semiconductor Industry Evolution

My View of EDA from the Top of Mentor Graphics

I joined Mentor Graphics (now Mentor, A Siemens Business), in late 1993. I knew Mentor was in trouble because I was a large customer for Mentor at TI. I looked forward to what I knew would be a real challenge—the rescue of an EDA company that had committed to a strategy that was likely to fail. My wife supported the move and thought that Portland would be a good place to raise our very young children. Plus, Jerry Junkins, the CEO of TI, made it clear to me that any succession to his role wouldn't happen for at least ten years because of his own career plans. Sadly, he was wrong about the length of his tenure at TI. Jerry passed away at the young age of 58 in 1996.

Mentor's bet on Version 8.0 had taken it from the #1 position in EDA to #3. Most software companies never recover from that type of decline. Yet, I came to Mentor with an optimistic view. After all, most companies that have failed product generations can quickly shift to other innovations they have on the shelf and re-generate their momentum. However, there wasn't a lot on the shelves to build upon, and almost everyone in the company had been moved to the Falcon project to try to save it.

We were able to stop the cash drain with some painful workforce reductions and a decrease in Version 8.0 spending that allowed us to find areas where we could be the de-facto standard. The shelves were not totally bare. For example, Mentor's system design business succeeded despite the difficulties of the Falcon Version 8.0 transition. Russ Henke, who managed the PCB business at that time, did not believe that Version 8.0 would ever work. So he followed a path, common in many companies, of quiet non-compliance. He instructed his PCB team to develop a "wrapper" to interface to Version 8.0, just in case it worked, and then proceeded to invest in the traditional PCB design business, consistently growing PCB revenue throughout the period of Version 8.0 chaos and into the 1990s.

The Mentor sales force had very little to sell after the announcement that Version 7.0 would not be extended but would be replaced by Version 8.0 whenever that environment became available. An innovative sales team working with the "Value Added Services" group sought out new users for the existing products that were not affected by the Version 8.0 transition. PCB schematic capture was one of those products. They found a local customer in Portland, Freightliner, who manufactured trucks and is now owned by Daimler. Convincing them to move from manual wiring design to EDA couldn't have been easy, but they became the first adopters of a "field-developed" product named "LCable," a name that reflected its use in the design and verification of cabling and wire harnesses for trucks and cars. Adoption by other automotive and aerospace companies proceeded slowly but, over the decade starting in 1992, the complexity of automotive and aerospace electronics increased so much that the need for EDA became apparent.

Developing Mentor's Strengths

When I started looking for the point-tools that Mentor should focus on, I first looked at their emulation. After all, I knew that Mentor had the best emulation technology in the industry, having visited there to observe it before. When I arrived at Mentor and asked about it, everyone started checking his shoe polish. Unfortunately, Mentor had sold its leading emulation technology, along with the patents, to Quickturn Design Systems, leaving only a very limited ability to compete.

I then turned to physical verification. After all, I had signed the contracts for Mentor to OEM TI's physical verification software while I was at TI, and it had been a reasonable recovery from Mentor's loss of Dracula (their OEM solution) when Cadence acquired ECAD and terminated Mentor's OEM agreement. TI was not interested in extending the OEM arrangement with Mentor to the next generation, so we bought out their rights and in January 1994, we had a big kick-off meeting to develop the next-generation of physical verification, headed by Laurence Grodd for the physical verification and Koby Kresh for the logic-to-schematic verification.

In addition to the fact that Laurence was brilliant, we had the benefit that he had maintained a database of designs that were verified using "Checkmate," the Mentor name for the product we OEM'd from TI. Laurence could handle hundreds of variants in design style. He proceeded to innovate innumerable approaches to physical verification including selective promotion and other things that are routine today. Unfortunately, Mentor didn't file any patents. So

ISS, a company in North Carolina that was ultimately acquired by Avant!, adopted many of these approaches, including hierarchical forms of analysis.

Internal politics were also a factor, as they always are in large companies. Mentor's custom IC layout product, IC Station, was in a battle to beat Cadence's product, Virtuoso. Our physical verification capability in IC Station, called IC Verify, came from Laurence and was clearly superior to the competition. So why would we sell it stand-alone to competitors using Virtuoso? Subsequently, a copy of "Calibre" was sneaked out to AMD and their designers became excited by it.

Meanwhile, Mentor's products that had evolved from the TI OEM continued to develop, and Intel became a major customer for a physical extraction product called MaskPE. MaskPE had been developed using TI's custom-design Checkmate architecture as its base, and Mentor was now repositioned and ready to compete with Checkmate. The game was definitely on.

At the next Design Automation Conference (DAC), a decision was made to also display the new "Calibre" capability, which appeared to undercut the roadmap that Intel was expecting for MaskPE. While the Intel surprise was upsetting for some in the Mentor sales force, Calibre clearly ushered in a whole new generation of physical verification and the name "Calibre" suggested that "ExCalibre" would be on the way to solve the physical extraction problem.

The critical role at this time was executed by Brian Derrick, GM of the Physical Verification Division. Brian did something very innovative, and probably forbidden in most large companies—he worked directly with Danny Perng, a salesman in Taiwan who was interested in focusing on Calibre for the foundries TSMC and UMC. Because our sales force knew that TSMC and UMC wouldn't pay much for tools, the sales and support resources in Taiwan were insufficient to drive a foundry campaign. So without permission, Brian hired his own sales force to complement Danny's effort. These specialists from the product division were able to convince the TSMC engineers, and later those at UMC, GLOBALFOUNDRIES, etc., that Calibre was superior to competitive approaches.

Simultaneously, Brian's team concluded that optical proximity correction (OPC) would be the next important extension of physical verification. Presim, a startup based in Portland, Oregon, was the leader in OPC, and they had captured the Intel account. Not to be defeated, Brian found the leading experts in the technology (going to UC. Berkeley to find OPC Technology Inc. and hiring Nick Cobb to head up the development). These strategic moves created the basis for Mentor's #1 position today in both physical verification and resolution enhancement.

As things progressed, Mentor had a lot of strong, even best-in-class tools: Calibre physical verification, Tessent design-for-test, Expedition PCB design, Calypto/Catapult high-level synthesis, automotive embedded electronics, and eight others, by the metric provided in the official Gary Smith EDA analyses. Fortunately, Synopsys eventually decided that they didn't have to do everything; they could pursue new areas that Mentor was not pursuing. That allowed a level of diversification that had not been common in the EDA industry.

And, with that, the EDA industry started to change. Each major EDA company developed specialties, instead of spending all its time trying to take market share from the others. And they all became more innovative. If I could claim one contribution to the EDA industry, it would be this. We are now an industry that looks for capabilities that will help our customers, and then develops (or acquires) those capabilities, rather than just trying to take market share from each other.

By the year 2000, the business was blossoming and had outgrown its original roots in PCB design and layout. Martin O'Brien joined Mentor from Raychem and brought with him a detailed knowledge of how automotive, aerospace, and defense companies thought about electrical wiring architectures. That became one of the valuable core businesses of Mentor over time. Today, the Capital family of integrated electrical system design products has become the leading system connectivity design environment, extending from concept through simulation, topology, bill of materials, factory form boards for manufacturing, and maintenance after the sale. Siemens has become a teaching customer but the Capital family is intensely focused on providing an open environment that can help Siemens' competitors as much as it helps Siemens.

A company in Mentor's position in the early 1990s rarely recovers, but Mentor was an exception. It pulled out of a tailspin by focusing on areas of specialization where it could be number one, and was saved by its variety of world-class point tools. That strategy worked, but it took a long time.

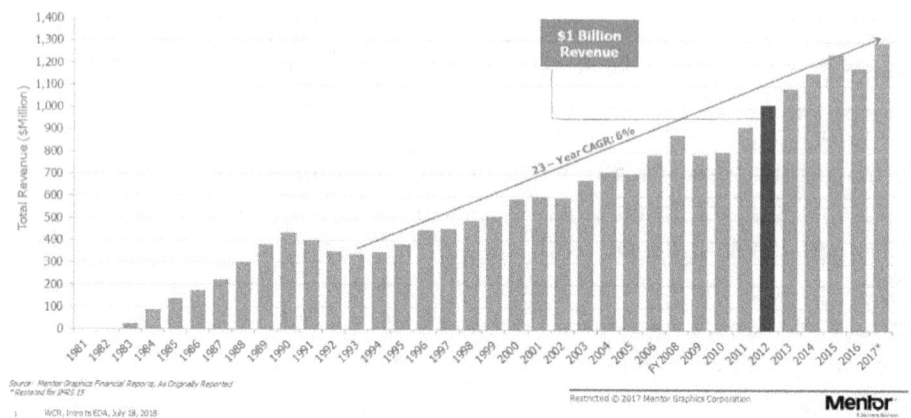

Mentor Graphics revenue.

Throughout this transition, the leading competitors adopted, and argued for, a new paradigm. That paradigm was a single-vendor flow, which also never evolved. Why? Because no one company can be the best at everything. The integration of tools and methodologies from different companies became critical to all those who wanted best-in-class design environments.

Today, Mentor's strategy of specialization still works. EDA is a business like the recording industry. There are rock stars, and they develop hits. Once a hit becomes entrenched, it's very hard to displace. Mentor focused on a few key areas where its position is hard to attack. Physical verification through the Calibre family is an example. Calibre is the golden-signoff tool, even though there are foundries that will grudgingly accept alternatives. When the debate about a variation in design rules occurs, the discussion between design and manufacturing people always returns to Calibre. PCB technology has similarities. You can use a variety of less expensive tools but why make life difficult for yourself?

Tessent design-for-test, or DFT, became a hallmark of this specialization strategy by putting together a group of the world's best test people and letting them do their thing. Under Janusz Rajski and a large group of test gurus, unique technologies like test compression, cell-aware test, hierarchical test, etc. were developed and used to build a commanding market share. Other areas where this point tool strategy was used to grow a complete design platform included high-level synthesis, optical proximity correction, automotive wiring, and others for a total of thirteen out of the forty largest segments of EDA, according to Gary Smith EDA.

Of course, there were many more battles to win (and lots of fun yet to be experienced). Whenever I ask successful people in technology, including CEOs, about the most enjoyable part of their careers, they almost always point to a period when they worked with a group that overcame the impossible and developed a product or capability that changed an industry. Calibre provided just such an experience for me and many others, as did a number of other developments that emerged on Mentor's path to recovery.

Tales from Mentor CEO Seat—Avant! plays the Acquisition Game

The departure of Gerry Hsu from Cadence in 1994 to form Avant! (originally named ArcSys) is chronicled in legal testimony as accusations of theft of software were followed by legal battles, financial awards, and even prison terms. Mentor and Synopsys were simply onlookers as the drama unfolded, but both had an interest in the outcome. The outcome of the trial pointed to substantial civil damages that Avant! would have to pay to Cadence. Mentor went to work with some of the top legal advisors at O'Melvany and Myers to estimate just how much those damages would be. Synopsys was reluctant to engage but was worried that, if Mentor acquired Avant!, the EDA balance of power could shift.

Gerry Hsu took up residence in Taipei, having avoided criminal charges for which some of his employees were not so lucky. Chi-Foon Chan, then EVP of Synopsys, suspected that Mentor was negotiating with Gerry Hsu to buy Avant! and Chi-Foon has since told me that he called every major hotel in Taipei to see if I was registered as a guest. In reality, we were much more serious about buying Avant! than Chi-Foon imagined. I rented an apartment in Taipei and spent more than a month living there and regularly meeting with Gerry. Meanwhile, Greg Hinckley, who was then Mentor CFO but effectively becoming

COO, conducted meetings with the investment bankers to determine how we could put together a successful proposal to buy Avant!.

The bankers paid a lot of attention to two issues: 1) Negotiating how much they would be paid for the transaction and 2) Removing the absurd benefit in Gerry Hsu's contract as Avant! CEO that would pay him $10 million if he left Avant! for any reason. Why would a Board of Directors approve such a condition? Avant!'s board at that time consisted of five people, four of whom were employees who reported to Gerry, and the fifth was a forestry major whose knowledge of semiconductors and EDA was very limited. Securing approval for this condition couldn't have been very difficult for Gerry even though it seemed to stand in the face of most responsible corporate governance. Greg, who is one of the best "out-of-the-box" thinkers I've ever known, addressed the bankers with a different question. "Why don't we triple the amount," suggested Greg, "and offer to pay Gerry $30 million instead of $10 million?" The bankers were aghast. Why would we do that? Greg's response: "There's obviously only one decision maker for the sale of the company so why don't we appeal to his self- interest?" The bankers were skeptical but we put together a proposal that incorporated this feature. As justification, we asked that Gerry extend his non-compete agreement from one year to three years in trade for tripling the severance payment.

I arranged to have dinner with Gerry in Taipei. He brought his son along and I presented the proposal. When I highlighted the change in severance arrangement for Gerry, he quickly became suspicious and began arguing with me that he was entitled to the $10 million severance payment. I had to repeat twice that I didn't dispute his right to the payment; I just wanted to extend his non-compete agreement to three years and triple the severance payment. Once Gerry understood, he became enthusiastic about the proposal and asked how quickly we could close an agreement. I cautioned Gerry that the terms of the agreement must be confidential, and I had Gerry approve the letter of intent and confidentiality agreement. We shook hands on the deal and I called the Mentor team to join us in Taipei to finalize the agreement.

I can't be sure how Gerry communicated with Synopsys but, by the time the Mentor negotiating team arrived, Gerry was already expressing second thoughts about his agreement to be purchased by Mentor. It became apparent that he was talking to another potential buyer despite his commitment to Mentor. So we returned to the U.S. with no deal.

Subsequently, Gerry's team contacted our bankers to re-start negotiations but we held firm, responding that we didn't feel we could trust him based upon our previous experience. We didn't engage again. Negotiations between Synopsys and Avant! continued and a deal was announced on December 3, 2001. A long period of review by the International Trade Commission ensued. After more than six months, the transaction was approved. Details were then published in a joint S4A filing by Synopsys and Avant![14] Among the most interesting details for me were:

1. Synopsys hired attorneys to estimate the cost of the civil damage award that would likely be incurred, just as Mentor had done, and the answer came out nearly the same as the estimate that Mentor had received. This was somewhat remarkable when you consider the uncertainty of outcomes in the U.S. legal system for disputes in high technology.

2. The agreement between Synopsys and Avant! included a $30.6 million cash payment to Gerry Hsu for his employment agreement. He didn't ever thank me.

There was a benefit for Mentor, however. Cirrus Logic was one of the first to detect anomalies in the Avant! software that led them to believe that the Cadence accusation of theft was credible. Under certain conditions, wavy lines appeared on the screen with the Avant! place and route software in the same manner as Cirrus had experienced with Cadence place and route software.

Mike Hackworth, CEO of Cirrus Logic, became concerned and talked with Joe Costello, CEO of Cadence, about switching back to Cadence for place and route. Mike's limitation was that Cadence would have to develop a tighter integration with Mentor's Calibre design rule checking software, which Cirrus had adopted. Mike, Joe, and I had a three-way conference call where I insisted that we needed to obtain detailed specifications for Cadence's LEF and DEF standards. Joe readily committed and assigned Bob Wiederhold, previously CEO of HLD, a company that had been acquired by Cadence, to effect the transfer of information. That's when we found out that DEF was not one standard, even in Cadence. There were many versions and interpretations. Despite all this, we were able to work together, and Calibre became tightly integrated with Cadence, and also Synopsys, making it successful in most of the design flows in the industry.

Tales from Mentor CEO Seat—Carl Icahn Comes Knocking

Carl Icahn is a remarkably charming person. You might expect him to be a mean, aggressive adversary but he actually jokes about his foibles, tells stories about interesting people, and gently poses thoughtful questions. "I thought Jerry Yang just didn't want to sell his Yahoo baby to Microsoft," Icahn related. "So I bought a few hundred million of Yahoo stock and called Steve Balmer, telling him we could make a deal. Steve said Microsoft had moved on. And you know, after my tenth call to him, I began to think they really had moved on," quipped Icahn. This seemed to relax some of the tension in the room, but I remembered my rehearsed preparation for the meeting.

There is an entire cottage industry of consultants who train executives in the art of dealing with Icahn. Mine was a day of training from one of the best firms, plus lots of study. More than 25 M.S. and Ph.D. theses have been written analyzing Icahn's tactics. Unlike Jeff Smith of Starboard and Jesse Cohn of Eliot Associates, both of whom I've dealt with, Icahn is unique. Less analytics and lots of gut feel.

Before entering Icahn's office, I knew what the room would look like, where I would be asked to sit (with the sun shining in my eyes), how he would start the conversation, what he would try to establish during the meeting and exactly what I should try to achieve. The year was 2010, and Icahn Associates had acquired over 10% of the common stock of Mentor Graphics. They planned to continue buying but were stopped by our "poison pill" that limited them to a 15% ownership. Donald Drapkin of Casablanca Capital followed Icahn's lead and began acquiring Mentor stock as well as appearing on television, as Icahn was doing, to blast the Mentor management.

And then the proxy fight followed, with three nominees from Icahn Associates to replace the most senior Mentor Directors. There's nothing like a proxy fight to consume time, upset employees and customers, and challenge the patience of a CEO. Every word and every slide that the company management communicates to anyone must be publicly disclosed in an SEC filing the next day. And each of these will be scrutinized for absolute accuracy. On the other side, the activist is free to make baseless accusations, misrepresent facts and generally stimulate unrest among shareholders and the public. Rules for a proxy fight clearly favor the activist and are not likely to be changed. The company is legally prohibited (in our case by court injunction) from explaining to shareholders how to split their ballots if they want to vote for less than all the proposed nominees of the activist. The result: Companies frequently negotiate a compromise with the activist, adding one or more activist-sponsored directors to their list of nominees. Some, like Mentor, fight the good fight but usually lose, as we did.

Then the challenge begins of managing a company when new directors will vote against most things that management proposes. In addition, much of the effort of the company is now directed at providing analyses for whatever objective the activist is promoting. In our case, that was the idea that Mentor should be sold or, at the very least, split into pieces to facilitate a sale.

And then there are the "shareholder" lawsuits that follow. Mentor spent hundreds of thousands of dollars defending a shareholder lawsuit claiming that we had improperly turned down an offer (which was actually not an offer) to buy the company for $18 per share. Through most of the years that the lawsuit continued, with depositions of the Mentor Directors and much of management, the stock was selling for more than $18 per share. If we lost, I wondered if the shareholders who were supposedly harmed would be required to pay us the difference between the $18 per share and the $20+ per share that their stock was now worth.

In most cases I've observed, the new board members begin to understand over time why the other board members and management have made the decisions they have made. Divergent director opinions gradually begin to converge. At the next Christmas after the proxy fight, I received an engraved bottle of Johnny Walker Blue scotch from Carl with the words, "NOT FOR USE AT BOARD MEETINGS." At a subsequent Christmas, after our stock price had increased substantially, I received one that said, "TO BE USED AT BOARD MEETINGS." Of course, I had to donate the bottles to charities or pay compensation to the company to avoid questionable receipt of a gift.

Gifts from Carl Icahn.

For Mentor and Icahn Associates, the ending was good. The Icahn stock appreciated from a purchase price near $9 to a peak of over $25, and Icahn Associates more than doubled its investment when Mentor bought back half the stock at $18.50 and Icahn sold the rest. We discovered things about our financial and business structure that we might not have investigated if we had not been stimulated by our new director demands. Although two of the three Icahn directors were not re-elected, the other one, David Schecter, was a strong contributor to the Board and we were sorry when he resigned.

The lesson for companies that come under attack? Continue to do what is best for your shareholders and resist acting in the interest of a minority shareholder just to reduce the pain of conflict. And keep an open mind; many of the themes that activists promote have merit even if they are driven by incomplete information. Ultimately, we all have the goal of increasing shareholder value and smart people working toward the same goal can usually find common ground.

EDA Cost and Pricing

When I moved from the semiconductor industry to Mentor in 1993, I expected most of my technology and business experience to apply similarly to EDA software. To some extent, that was correct. But there was a fundamental difference that required a change in thinking. Product inventory, especially for semiconductors, must be minimized because it has both real and accounting value. We used to say that semiconductor finished goods are like fish; if you keep them too long, they begin to smell. Software inventory doesn't even exist. When the order is placed, the actual copy of the software is quickly generated and shipped.

EDA customers are aware of this "software inventory" phenomenon. There is no deadline for purchases that takes into account the lead time for manufacturing, as there is when ordering semiconductor components; an EDA order placed near the end of a quarter can be filled within that quarter. Negotiations for large EDA software purchase commitments tend to drag on until near the end of the quarter when customers suspect they will get the best deal. To counter that, EDA companies provide incentives, or other approaches, to minimize the last minute pressure.

If negotiated terms are not satisfactory, why don't the EDA companies just let the orders slide into the next quarter? Sometimes they do. But, unlike the semiconductor industry (or any other manufacturing industry), EDA software is between 90% and 100% gross margin. Profit is therefore asymmetric. In a semiconductor business, a dollar of cost reduction improves profit by one dollar and a dollar of incremental revenue increases profit by typically 35 to 55 cents. So, cost reduction for a semiconductor company has a larger impact on profit percent than revenue growth.

EDA Market Dynamics

One great aspect of the EDA industry is the ability of new startups to successfully introduce a new point tool and grow to be valuable enterprises. Most of these companies were acquired by one of the "Big Three." Since the mid-1970s, the EDA market share has been dominated by three companies at a time. Computervision, Calma, and Applicon gave way to Daisy, Mentor, and Valid and then Mentor, Cadence, and Synopsys.

It seems that three large EDA companies is a stable configuration as long as technology is evolving rapidly. One reason for this phenomenon is that EDA tools are very "sticky"; there is typically a de facto industry standard for each specialization, like logic synthesis, physical verification, design for test, etc. I suspect that most EDA companies make most of their profit from the software products where they are number one in a design segment (GSEDA analyzed nearly seventy such segments for the industry that are over $1 million per year in revenue).

It's very hard to spend enough on R&D and marketing to displace the number one provider of each of these types of software, so changes occur mostly when there are technical discontinuities that force adoption of new methodologies and design tools. When a major discontinuity occurs, new companies will appear and we'll probably have another shakeout.

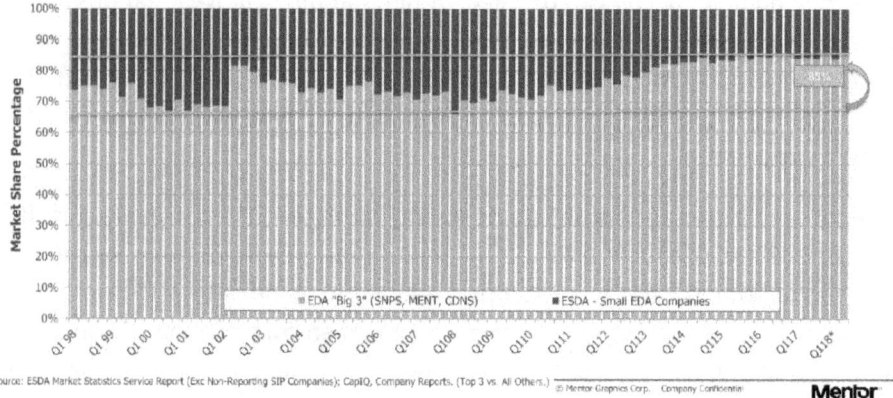

Market share of the Mentor, Cadence, and Synopsys from 1998 until 2018.

Interestingly, the combined market share of Mentor, Cadence, and Synopsys has remained almost constant over twenty years at 75% plus or minus 8%, despite all the acquisitions. Cadence grew almost exclusively by acquisition while Mentor did very few. Synopsys was somewhere in between. I once joked to a group of Cadence employees that, other than Spectre, I couldn't think of a single successful Cadence product that was conceived and developed within

Cadence. The group looked shocked and told me that I was not correct. "Every line of Spectre," they said, "was developed at U.C. Berkeley."

In the EDA industry (or most software businesses), a dollar of cost reduction has nearly the same profit impact as a dollar of incremental revenue. The conclusion: working on incremental revenue growth is just as productive, and a lot more pleasant, than working on cost reduction. In addition, a 5% miss in revenue produces about a 25% miss in profit (if the company's operating profit is normally 20%) so it's very disturbing to shareholders when an EDA company misses its revenue forecast because the accompanying profit miss will be large.

During the last decade or so, another profit leverage phenomenon has been important. With interest rates at very low levels, acquisition of another profitable company using cash is very accretive to earnings, i.e. cash that is sitting on the balance sheet collecting very little interest now becomes a profit-generating asset that increases overall earnings for the acquiring company. A variety of industry analysts and investors have even taken the position that a company that doesn't fully utilize its borrowing power is under-utilizing a corporate asset. That ignores the risk element associated with borrowing, but it has encouraged lots of acquisitions, especially in high technology.

Although there has been some consolidation in the semiconductor industry, the primary change has been a higher degree of specialization. Companies like TI that once produced nearly every type of semiconductor and very little profit now produce primarily analog and power devices and consistently deliver among the highest profit in the semiconductor industry. Similarly, NXP moved from a broad mix of products to specialization in automotive and security components.

Finally, I find it interesting to look at the EDA cost from the perspective of the companies purchasing the software. Frequently, I hear the complaint that EDA companies never lower the price of their software while semiconductor companies have to reduce the cost per transistor of their chips by more than 30% per year. To analyze this complaint, we used published data to look at the "learning curves" for semiconductors and EDA software.

From Wild West to Modern Life: Semiconductor Industry Evolution

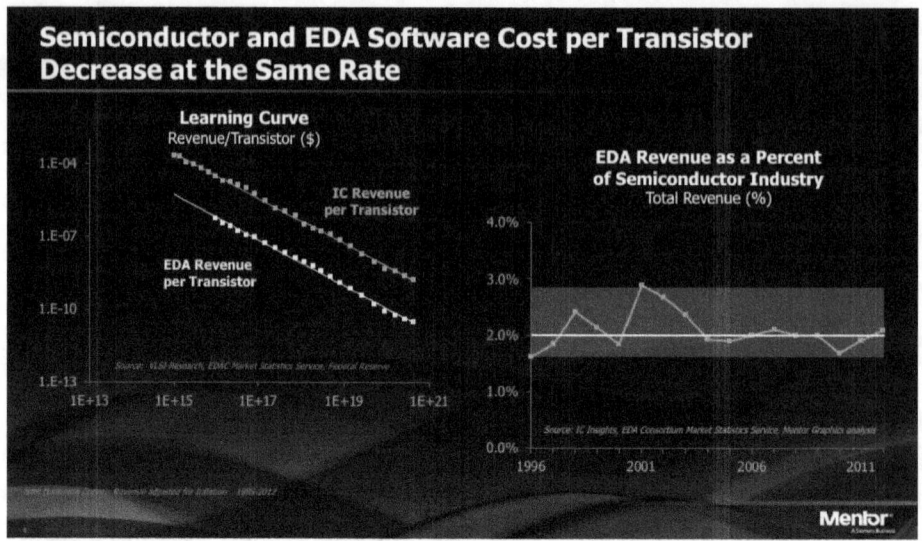

EDA cost per transistor.

A semiconductor learning curve is a plot of the log of cost (or price) per transistor on the vertical axis and the log of the cumulative number of transistors ever produced on the horizontal axis. For free markets like semiconductors, the result is a straight line, presumably forever. Semiconductor industry analysts publish data on the number of transistors produced each year as well as the revenue of the industry. The Electronic System Design Alliance publishes the revenue of the EDA industry.

When you plot the semiconductor industry revenue per transistor and the EDA industry revenue per transistor on a learning curve, as in the figure, the straight lines are exactly parallel. That means that the EDA software cost per transistor is decreasing at the same rate as the semiconductor revenue per transistor. Is that a surprise? It shouldn't be. Just as the semiconductor industry has spent 14% of its revenue on R&D for more than the last thirty years, it has spent 2% of its revenue on EDA software for nearly twenty-five years. If EDA companies failed to reduce their price per transistor at the same rate that the semiconductor industry must reduce its cost per transistor, then EDA would become a larger percentage of revenue for the semiconductor companies, exceeding the 2% average and forcing reduction of other semiconductor input costs.

It turns out that the total semiconductor ecosystem behaves similarly, reducing costs to provide better products and a six order of magnitude decrease in the cost per transistor over the last thirty-five years.

Honey, I Shrunk the EDA TAM

Throughout the history of the EDA industry, pricing models have caused discontinuities in the way the industry operates. For a variety of competitive reasons, individual companies have developed ways to change the pricing model in an attempt to secure a competitive advantage. Following are some of the most memorable:

Valid Logic (1988) – Remove the premium for "global float" and allow all licenses to "float" around the world. This one sounds pretty reasonable in today's computing server environment, but in 1988, software licenses were "node locked." You purchased a design software license for one work station and it could "float" only within a reasonable distance, say around a single corporate site. Valid Logic offered their customers free float of the license to any of the customer's worldwide locations through a program called "ACCESS." It was a big hit. It also destroyed a significant portion of the total available market for EDA software, more than half by some estimates, as other EDA companies followed suit.

Avant! Subscription Licensing – In the mid-1990s, Avant! introduced a three-year time-based licensing model. I am told by Daniel Nenni it was driven by Avant!'s observation that customers purchased perpetual licenses that lasted for about three years (two Moore's Law process nodes) before they had to upgrade and buy new perpetual licenses (although Red Herring magazine reported that Gerry told them he got the idea from car leasing plans). At this time, the industry model was a combination of perpetual licenses plus ongoing maintenance. The maintenance fee was 15-20% annually of the cost of the perpetual license, similar to what most of the non-EDA software industry offers today, except for the more recent introduction of SAAS (software as a service) models. The perpetual license cost was high and the revenue was all recognized "up front" because the customer now owned the software. For Avant!'s three-year subscription model, the entire EDA industry followed the example like lemmings because of pressure from customers. It also had an attraction for the

EDA companies since it offered a continuing revenue stream and EDA companies were worried about what would happen when perpetual license sales slowed to a smaller percentage of their revenue and maintenance revenue became the primary ongoing revenue source. The problem with the three-year subscription model was that competitive discounting quickly drove the subscription price down to about the same level as the previous annual maintenance cost. Now the customers were receiving product plus maintenance for the same cost as they previously paid just for maintenance, a good deal for the customers but questionable for the EDA companies.

Cadence FAM (Flexible Access Model) – This was introduced in the late 1990s. It was essentially a three year "all you can eat" approach to software from a single EDA company. It was a hit with the Cadence sales force and the customers but it caused lots of disruption in the industry although I don't think other companies offered anything similar. It led to internal management disruption at Cadence. At the Cadence earnings call on April 20, 1999, the company announced that "the company has run into a 'one to two-quarter delay in absorption of 0.18-micron design tools' among semiconductor makers." Many in the EDA industry translated this as: "A large number of our best customers have purchased three-year FAM licenses so we can't collect additional revenue from them for a while."

Cadence Re-Mix – Once again, Cadence set the pace of innovation in pricing with the introduction of "Re-Mix." A customer specifies the mix of software products desired on the date of contract renewal but, if the customer chooses to change the relative mix of one product versus another, he can do so within the limits of the original contract value. Up until this time, customers had to guess what their mix of product needs would be for the next three years. Typically, they had to buy twice as much software as they would use on an ongoing basis because they couldn't predict the mix of products they would need. The result: By some estimates, this re-mix approach eliminated as much as half of the EDA total addressable market (TAM) because customers didn't have to predict their future mix of needs and didn't have to buy licenses sufficient for peak usage.

Foundry IP libraries – Until the late 1990s, silicon foundries like TSMC left the entire design process to their customers. TSMC received a verified GDSII file from the customer and they checked it and then generated photomasks, fabricated wafers, and shipped parts to the customer. Companies like Artisan were in the business of creating physical libraries of standard cell blocks that were checked for correctness and modeling by being fabricated on a test wafer by the foundry. The libraries were then sold to customers doing the designs to speed design of the standard, undifferentiated parts of their chips. Wouldn't it be great if customers could have access to the entire Artisan library during the design phase and then only be charged based upon the number of cells that were actually used in their designs multiplied by the number of chips produced? Artisan thought so. And they convinced TSMC to adopt the model, providing software to trace the usage of Artisan cells. Artisan consequently developed a stable stream of royalty revenue from TSMC, making them an attractive acquisition for ARM. I'm told that the deal was not so good for TSMC. High-volume customers negotiated discounts to wafer pricing with TSMC and the standard cell libraries became part of those negotiations. As a result, the additional money that TSMC expected to receive from their customers by charging them for the use of standard cells turned out to be elusive. The bundled price of wafers plus photomasks plus IP, etc. was included in the wafer price and any incremental revenue for the cell libraries was hard to find.

You may wonder how I can be so cavalier about this whole topic when, during the last twenty-five years as CEO of Mentor, my company was subjected to so much cost and revenue pressure by these model changes? Because the revenue of the EDA industry has continued to be 2% of semiconductor revenue for more than twenty years. These model changes were simply part of the way that discounts were provided to customers so that the EDA companies could stay on the learning curve and give semiconductor companies a reduced cost per transistor for design software. If the pricing models hadn't changed, we would have had to provide those discounts in some other form because the EDA industry had to reduce its software price per transistor at the same rate that the semiconductor industry reduced its revenue per transistor.

Basic Techniques for Managing an EDA Business

Starting in 1997, I had the privilege to work with Greg Hinckley, a superb "out of the box" thinker and excellent operational manager. We had such a common view of the world that we almost always agreed upon the best approach to problems and opportunities and hardly needed to confer, although we regularly did discuss almost all significant issues. One of our common attributes was a contrarian attitude. If everyone else was headed in the same direction, opportunity must lie in a different direction. That belief in "do what others aren't doing" served as a guidepost for our moves at Mentor.

Following are some of the approaches that Greg or I brought to Mentor, or that evolved from combining our previous experience through interaction together.

- Listen for silence
    - Divisions, product groups, people and managers are always vying for the attention of top management and will usually let you know the good things they are doing
    - If you never hear from, or about, an entity in your organization, there are two possibilities:
        - They are doing nothing worthwhile and make no difference to the company, in which case they should be eliminated, or
        - They are trying to hide something, in which case they should be investigated
- Life never gets any easier
    - Ignoring the problems of today only forces you to deal with bigger problems tomorrow
    - Never hope that excess inventory will be any easier to sell in the future than it is today—write it off now
    - Never assume that an overdue account will find a way to pay you in the future—it's even less likely than the present
- What you do infrequently, you do poorly
    - Send an annual invoice, not a quarterly one, for your customers to renew their maintenance contract—the person who deals with it will have to depend upon you to tell him how to handle it

- Restrict the unfavorable options in your contracts so that customers only deal with them at renewal time. If they have to deal with options regularly, they will become very good at it.
    - Don't expect your own organization to become efficient at things they rarely deal with—try to automate those things or consolidate them with someone who deals with similar issues regularly
- Making rational decisions based upon data is more important than being right all the time
    - Can you provide a rational basis for why you are making the decision?
    - If you turn out to be wrong, you can retest your rationale and make a better decision next time
- Budgeting starts at the top
    - Asking organizations to determine how much they need to spend in the coming year (or quarter, month, day, etc.) is a worthless exercise and a waste of company resources—the answer will almost always be more than the company can afford
    - Budgeting must start with an objective sales and revenue decision, heavily influenced by those who will have to achieve the sales. Budgets can then be developed that will fit into that sales and revenue plan.
    - Never be tempted to increase the revenue forecast when the cost budgeting pressures appear insurmountable. It will be even more difficult to reduce the budget later when the revenue comes in at (or below) the realistic level that is lower than the optimistic revised forecast.
- In a chip company, management needs to focus first on what is NOT working. In a software company, management should focus first on what IS working.
    - Problems in a semiconductor company (low yields, rapidly falling prices, excess inventory, etc.) can bankrupt you quickly; fixing them has immediate, measurable benefit and improves profit with cost reduction, dollar for dollar
    - In a software company, inventory is unimportant and incremental revenue nearly equals incremental pre-tax profit. Costs are largely fixed because they are mostly people-related and people should not be treated as a variable expense. Incremental revenue is everything so

- maximum effort should be focused on using existing resources to produce additional revenue.
- Spend your time doing what others are NOT doing
  - Strategies that are similar to what the other mainstream companies are doing will rarely develop any unique advantage for your customers
  - Needs that are not being addressed offer more opportunity than those that are receiving lots of attention
- For an employee, it is a great success for management to adopt your idea and claim it as its own, even if you are not acknowledged as the creator. For management, it is a great success to see employees act upon your idea and claim it as their own, even if they do not acknowledge you as the source.
- In technical software, users rarely change the product they are using until it becomes incapable of completing the required task—not when it becomes, slow, inefficient, hard to use or unpopular.
- Benchmarking with competitive products results primarily in price competition for the customer to secure a lower price to continue using the same software
- Companies that offer a new and improved product would be better off to use their time by:
  - Participating in benchmarks of new applications that can't be handled by the older, competitive tool, or
  - Finding new applications for the tool where there is not already an incumbent supplier.
  - When you acquire an EDA competitor, you don't eliminate competition—you create it. In EDA, people are the competitors and they keep doing what they are good at doing by creating new companies whenever the one they are working for is acquired.
- A new product or service is not really profitable until you receive your first order from a customer with whom you've never interacted
  - As long as orders come only from direct effort and interaction by your own organization, you are expending too much money for the revenue you are achieving
  - When word of mouth and third-party support start generating orders, you've gone viral and have a winning, profitable product
- If you only play defense, your competitor will eventually score.
  - You must counterattack to make your competitor play defense

- Selling your new, innovative products to your existing customers may be a starting point but success comes only when you win new customers who are, usually, already served by your competitor
    - Statistically, your competitor will eventually get lucky if he keeps attacking your accounts and is not burdened with defense of his own accounts if you don't attack them.
- When your company is not number one in a specific market, always compare yourself to the leader(s). When, or if, you become number one, never compare yourself to anyone.
    - Market communication is all about consideration to purchase. You want to be considered when purchasing decisions are made, so the association with the leader enhances the possibility of customer comparison of the leader's product or service with yours.
    - If you are number one in a market, it's counterproductive to provide your customer with a list of competitors to evaluate, even if you think your product is superior.

This is not meant to be a complete formula for successful EDA management. It's a collection of ideas that aren't normally discussed in business management training. Yet I believe that these observations, combined with other, standard good-management practices and an unwavering commitment to ethical treatment of customers, employees, and shareholders, helped Mentor to recover from a difficult period and ultimately demonstrate leadership and success in a difficult high technology business.

References

[1] http://semiconductormuseum.com/Transistors/RCA/OralHistories/Pankove/Pankove_Page2.htm
[2] Stevenson David A, Rhines Walden C, Maruska Herbert P. Gallium nitride metal-semiconductor light-emitting diode; June 25, 1974 [U.S. Patent 3,819,974].
[3] H.P. Maruska, W.C. Rhines, D.A. Stevenson. Mater Res Bull, 7 (1972), p. 777
[4] A modern perspective on the history of semiconductor nitride blue light sources, https://www.sciencedirect.com/science/article/pii/S0038110115001318
[5] https://spectrum.ieee.org/tech-talk/geek-life/profiles/nakamura-gives-some-credit-to-maruska-for-blue-led-invention
[6] spectrum.ieee.org/tech-history/heroic-failures/the-inside-story-of-texas-instruments-biggest-blunder-the-tms9900-microprocessor
[7] The SR-52 cost $395 on release in 1975 which is roughly $1,847 in 2018
[8] https://news.stanford.edu/2005/06/14/jobs-061505/
[9] https://patents.google.com/patent/US3379584A/en
[10] U.S. Patent 5,061,049, Spatial Light Modulator and Method, Inventor L.J. Hornbeck
[11] https://spectrum.ieee.org/tech-history/heroic-failures/the-inside-story-of-texas-instruments-biggest-blunder-the-tms9900-microprocessor
[12] http://www.tihaa.org/historian/TMS32010-12.pdf
[13] https://spectrum.ieee.org/tech-history/heroic-failures/the-texas-instruments-994-worlds-first-16bit-computer
[14] https://www.sec.gov/Archives/edgar/data/883241/000095012302004502/0000950123-02-004502-index.htm

www.ingramcontent.com/pod-product-compliance
Lightning Source LLC
Chambersburg PA
CBHW072233170526
45158CB00002BA/882